AMERICA'S FINEST ROCKETS™ PRESENTS:

ATLAS-CENTAUR 16

Source Documents

The contents of this book are available for free online. This book is a compilation of reprints of several National Aeronautics Air and Space Administration (NASA) publications and figures, all relating to Atlas-Centaur 16 and its payload, OAO-II. They were downloaded from online sources, including the NASA Technical Reports Server (NTRS, online at *www.sti.nasa.gov*) and *Scribd.com*.

Credits

These reports are generally attributable to the National Aeronautics and Space Administration (NASA) but, when possible, authors of specific reports or sections thereof are also given. The pictures on the cover are courtesy NASA.

Intent

Some readers do not like to read on a computer screen and prefer reading a physical copy. However, they may not want to suffer the price and inconvenience of a pile of unbound pages. This printed and bound copy of this report is priced low, to make it available to as many readers as possible. The reprinting of this report is not intended to offend NASA or the authors of the report, any of whom should contact the publisher with any questions or concerns.

Alterations and New Content

Nothing has been omitted from the original pdf files, although some figures have been enlarged to show detail. The pages in the OAO-II Press Kit, Release No. 68-186K, have been renumbered for publication purposes, but no other changes have been made. However, some New Content has been added to this compilation: (1) the cover; (2) the preceding Title Page; (3) this Publication Statement; and (4) the Summary of the contents of this compilation on page (i).

Copyright

ISBN-13: 978-1544970059

America's Finest Rockets™

AMERICA'S FINEST ROCKETS™ PRESENTS:

Atlas-Centaur 16

SUMMARY

The Atlas-Centaur 16 launch vehicle (AC-16) was launched on December 7, 1968, with its payload, the second Orbiting Astronomical Observatory (OAO-II). AC-16 delivered OAO-II into a nearly circular Earth orbit at an altitude of 772 km using a direct (single burn) mode of ascent. The flight also included a Centaur retro-maneuver after spacecraft separation.

CONTENTS

NASA TECHNICAL MEMORANDUM

NASA TM X-1989

ATLAS-CENTAUR AC-16 FLIGHT PERFORMANCE EVALUATION FOR THE ORBITING ASTRONOMICAL OBSERVATORY OAO-II MISSION

Staff of Lewis Research Center
Cleveland, Ohio

NATIONAL AERONAUTICS AND SPACE ADMINISTRATION . WASHINGTON, D.C. . MAY 1970

I. SUMMARY

The Atlas-Centaur (AC-16) with the Orbiting Astronomical Observatory-II (OAO-II) spacecraft was successfully launched from the Eastern Test Range Complex 36B on December 7, 1968, at 0340:09 eastern standard time. The direct-ascent flight profile was considerably more lofted than those employed in previous Atlas-Centaur flights. The vehicle was programmed to a flight azimuth of 60° in order not to violate the range safety protective zone around Bermuda. To achieve this flight azimuth the Atlas vehicle roll program was 5 seconds longer than for previous Atlas-Centaur flights. After it cleared the Bermuda protective zone, the Centaur was yawed to the right to achieve the desired orbital inclination of 35° to the equator. All Atlas and Centaur systems performed properly, and the spacecraft was injected into the desired 772-kilometer (417-n-mi) near-circular Earth orbit at the desired inclination.

The AC-16 was the first vehicle in which the Centaur flight control system provided the Atlas with both rate and position signal during the Atlas sustainer phase of flight. The OAO nose fairing, flown for the first time on an Atlas-Centaur vehicle, adequately protected the spacecraft during ascent and was jettisoned successfully. Because of the increased weight of the payload and fairing, the Atlas holddown time on the launch pad was extended 1.76 seconds to achieve the desired minimum thrust-to-weight ratio of 1.2 at lift-off.

This report presents an evaluation of the Atlas-Centaur system in support of the OAO-II.

II. INTRODUCTION

by John J. Nieberding

The purpose of the Orbiting Astronomical Observatory-II (OAO-II) was to make precision telescopic measurements from above the Earth's atmosphere. Specific areas of interest were mapping and studying the emission and absorption characteristics of the Sun, stars, planets, nebulae, and interplanetary and interstellar media in the relatively unexplored ultraviolet region of the spectrum.

The primary objective of the AC-16 flight was to inject the OAO-II spacecraft into a circular Earth orbit at an altitude of 772 kilometers (417 n mi) and at an orbital inclination to the equator of 35°. The launch vehicle was also required to perform a retro-maneuver to minimize the periods the Centaur would be in view of the spacecraft optical sensors and to minimize contamination of the spacecraft from Centaur exhaust products.

The AC-16 launch vehicle and the launch vehicle/spacecraft integration effort to support the OAO program were under the direction of the Lewis Research Center. OAO-II was the second of four OAO satellites currently planned to be launched. OAO-I was successfully launched on April 8, 1966, by an Atlas-Agena launch vehicle. OAO-III and -IV are currently planned for 1970 and 1971 launches, respectively.

The AC-16 used a direct (single burn) mode of ascent to inject the spacecraft into the desired orbit. Because of the required orbital altitude of 772 kilometers (417 n mi), the flight profile was considerably steeper than on previous Atlas-Centaur flights. The vehicle was programmed to a flight azimuth of 60° in order to avoid the protective zone around Bermuda imposed by Range Safety at the Eastern Test Range. Then, after it cleared the Bermuda protective zone, the Centaur was yawed to the right to achieve the desired orbital inclination of 35° to the equator. This report evaluates the performance of the Atlas-Centaur launch vehicle, in support of OAO-II, from lift-off through space-craft separation and completion of the Centaur retromaneuver following spacecraft separation.

III. LAUNCH VEHICLE DESCRIPTION

by Eugene E. Coffey and Joseph A. Ziemianski

The Atlas-Centaur is a two-stage launch vehicle consisting of an Atlas first stage and a Centaur second stage connected by an interstage adapter. Both stages are 3.05 meters (10 ft) in diameter, and the composite vehicle is 41.8 meters (135 ft) in length. The vehicle weight at lift-off is approximately 147 255 kilograms (324 656 lbm). The basic structure of the Atlas and the Centaur stages utilizes thin-wall, pressure-stabilized, main propellant tank sections of monocoque construction. Figure III-1 shows the Atlas-Centaur lifting off with the OAO-II spacecraft.

The first-stage SLV-3C Atlas (fig. III-2) is 21.03 meters (69 ft) long. It is powered by an MA-5 propulsion system consisting of a booster engine with two thrust chambers and with a total thrust at sea level of 1494×10^3 newtons (336×10^3 lbf), a sustainer engine with a thrust at sea level of 258×10^3 newtons (58×10^3 lbf), and two vernier engines with a thrust at sea level of 2980 newtons (670 lbf) each.

All engines use liquid oxygen and RP-1 (kerosene) as propellants and are ignited prior to lift-off. The booster engine thrust chambers are gimbaled for pitch, yaw, and roll control during the booster engine phase of the flight. This phase is completed at booster engine cutoff, which occurs when the vehicle acceleration reaches about 5.7 g's, and the booster engine section is jettisoned 3.1 seconds later. The sustainer engine and the vernier engines continue to burn for the Atlas sustainer phase of the flight. During this phase, the sustainer engine is gimbaled for pitch and yaw control, while the vernier engines are gimbaled for roll control only. The sustainer and vernier engines fire until propellant depletion. The Atlas is severed from the Centaur by the firing of a shaped charge system located on the forward end of the interstage adapter. The firing of a retrorocket system then separates the Atlas - interstage adapter from the Centaur.

The Centaur second stage (fig. III-3) is about 9.1 meters (30 ft) long. It is a high-performance stage (specific impulse, 442 sec) powered by two RL10A-3-3 engines which generate a total sea-level thrust of approximately 133.45×10^3 newtons (30 000 lbf). These engines use liquid hydrogen and liquid oxygen as propellants. The Centaur main engines are gimbaled to provide pitch, yaw, and roll control during Centaur powered flight. Fourteen hydrogen peroxide engines, mounted on the aft periphery of the tank, provide various thrust levels for attitude control after Centaur main engine cutoff and for vehicle reorientation after spacecraft separation.

The Centaur hydrogen tank is shielded with four insulation panel sections, each 2.54 centimeters (1 in.) thick. Each section consists of a polyurethane-foam-filled honeycomb core, covered with fiber glass lamination. The nose fairing system (fig. III-4) is 12.2 meters (40 ft) long and consists of a jettisonable section (fiber glass honeycomb core nose fairing and metallic split fairing) and a fixed metallic fairing and fiber glass honeycomb core barrel section. It is used to provide an aerodynamic shield for the OAO-II spacecraft, for the Centaur guidance equipment, and for the Centaur electronic packages during ascent.

This nose fairing system flown for the first time on Atlas-Centaur is similar to the Surveyor nose fairing except it is 5.5 meters (18 ft) longer and uses a mechanical spring system instead of a gas thruster system to jettison the fairing. The fixed fairing and the barrel section remain with the Centaur after jettison of the nose fairing. The insulation panels are jettisoned during the Atlas sustainer phase, and the nose fairing is jettisoned shortly after Centaur main engine start. The OAO-II spacecraft is shown in figure III-5.

Figure III-1. – Atlas-Centaur lifting off with OAO-11.

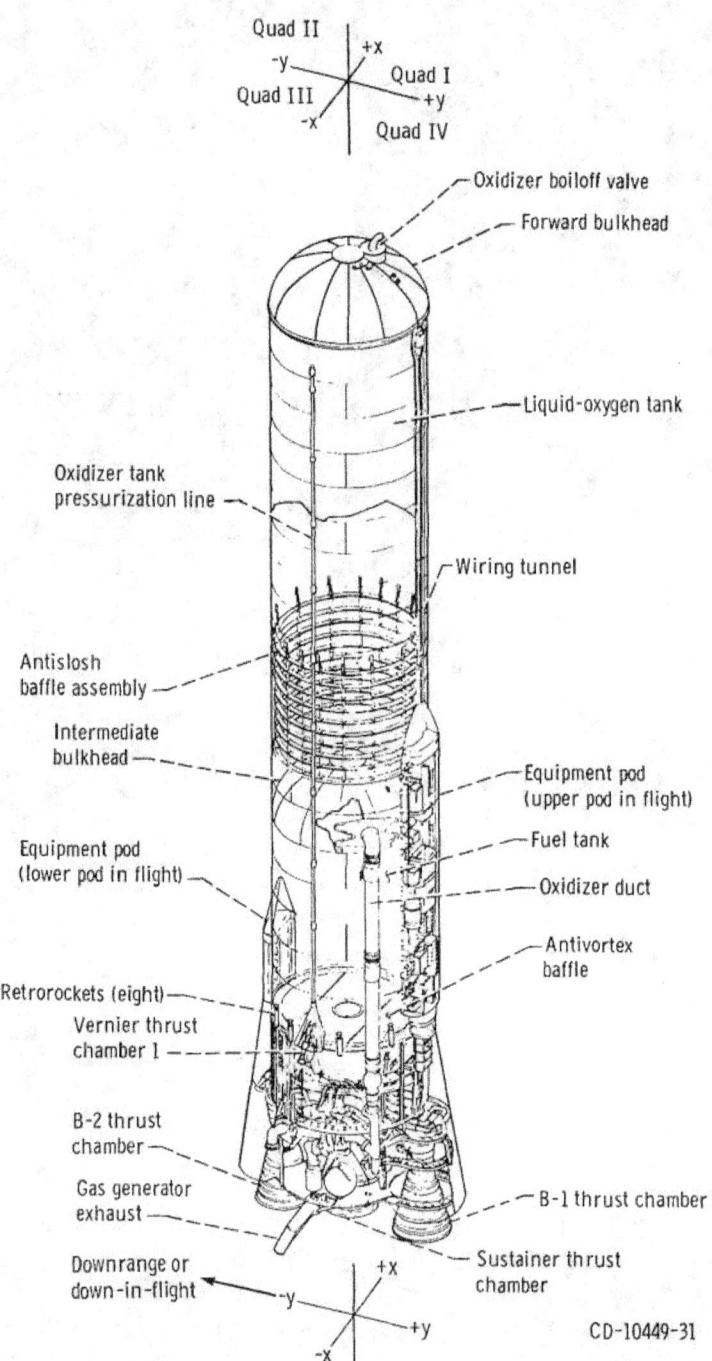

Quad II
+x
-y
Quad III
Quad I
+y
-x
Quad IV

Oxidizer boiloff valve

Forward bulkhead

Liquid-oxygen tank

Oxidizer tank pressurization line

Wiring tunnel

Antislosh baffle assembly

Intermediate bulkhead

Equipment pod (upper pod in flight)

Fuel tank

Oxidizer duct

Equipment pod (lower pod in flight)

Antivortex baffle

Retrorockets (eight)

Vernier thrust chamber 1

B-2 thrust chamber

Gas generator exhaust

B-1 thrust chamber

Sustainer thrust chamber

Downrange or down-in-flight
+x
-y
+y
-x

CD-10449-31

Figure III-2. - General arrangement of Atlas launch vehicle, AC-16.

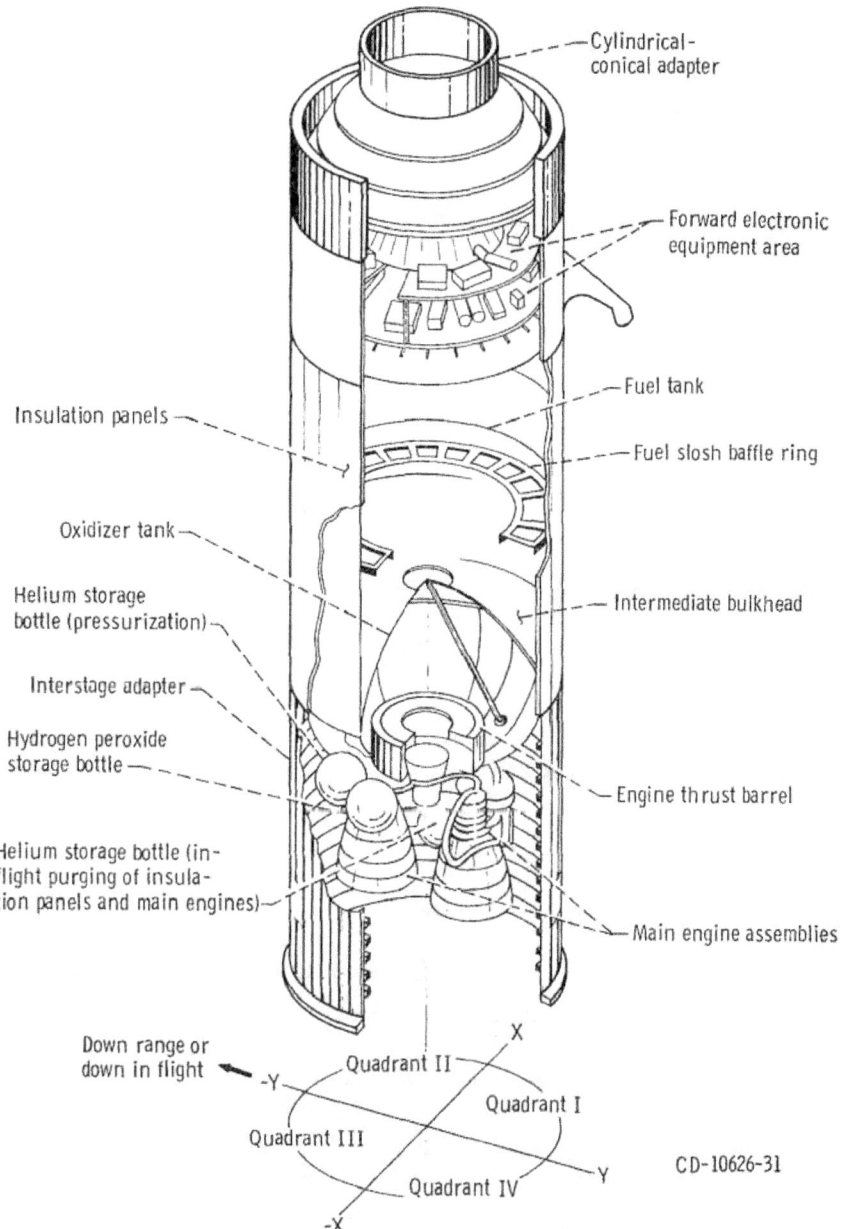

Cylindrical-conical adapter

Forward electronic equipment area

Fuel tank

Fuel slosh baffle ring

Insulation panels

Oxidizer tank

Intermediate bulkhead

Helium storage bottle (pressurization)

Interstage adapter

Hydrogen peroxide storage bottle

Engine thrust barrel

Helium storage bottle (in-flight purging of insulation panels and main engines)

Main engine assemblies

Down range or down in flight

X

Quadrant II

-Y

Quadrant I

Quadrant III

Y

Quadrant IV

-X

CD-10626-31

Figure III-3. - General arrangement of Centaur vehicle, AC-16.

9

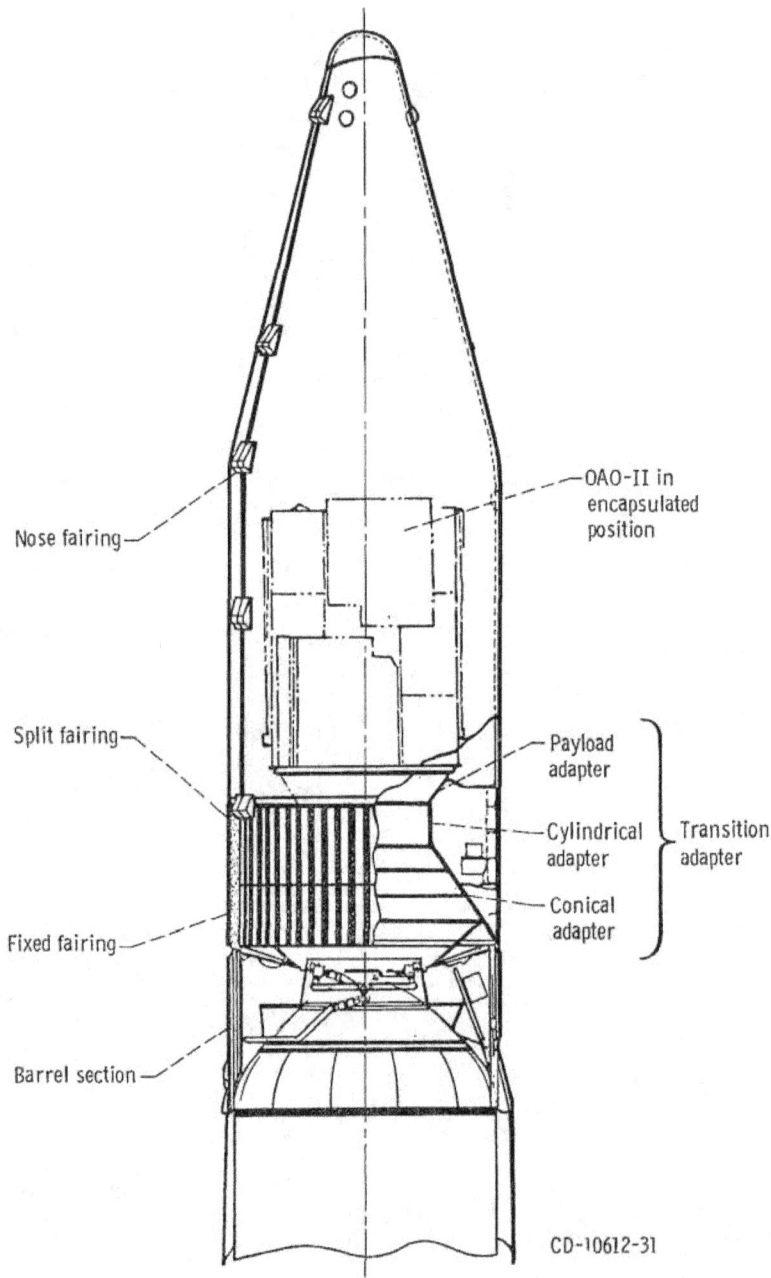

Nose fairing

OAO-II in
encapsulated
position

Split fairing

Payload
adapter

Cylindrical
adapter

Transition
adapter

Conical
adapter

Fixed fairing

Barrel section

CD-10612-31

Figure III-4. - Nose fairing system, AC-16.

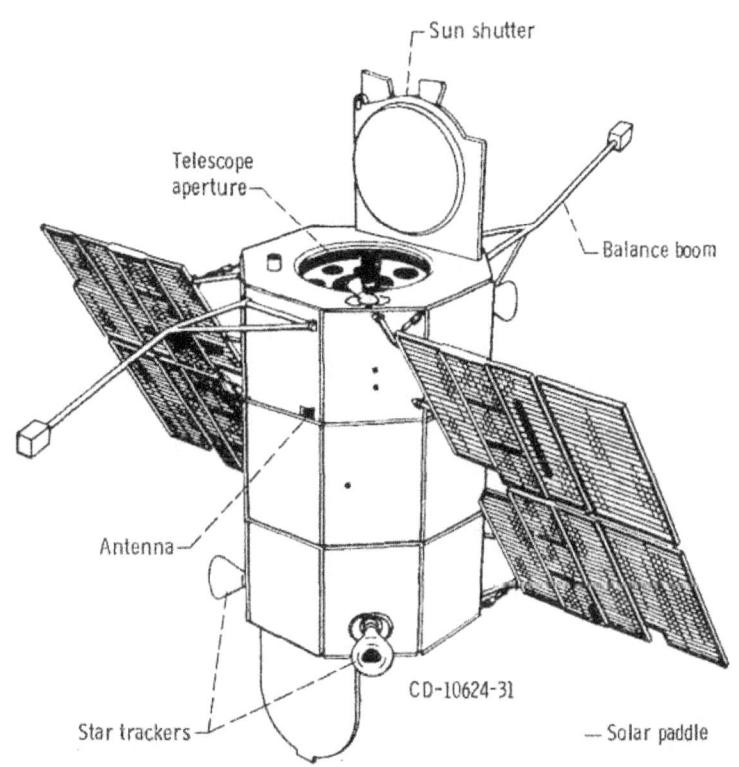

Figure III-5. - Deployed configuration of Orbiting Astronomical Observatory-II (OAO-II).

IV. MISSION PERFORMANCE

by William A. Groesbeck

The Atlas-Centaur vehicle AC-16, with the OAO-II spacecraft, was successfully launched from Eastern Test Range Complex 36B on December 7, 1968, at 0340:09 eastern Standard time. The launch vehicle used the direct (single burn) mode of ascent to place the OAO-II into a nearly circular orbit. All launch vehicle objectives were achieved, and the spacecraft was placed in Earth orbit at an altitude of about 772 kilometers (417 n mi).

The AC-16 mission profile and the OAO-II orbital trajectory are shown in figures IV-1 and IV-2. The OAO-II launch window and countdown history, as well as a summary of postflight vehicle weights, are given in appendix A.

ATLAS FLIGHT PHASE

Ignition and thrust buildup of the Atlas engines were normal. The vehicle lifted off (T + 0 sec) with a combined vehicle weight of 147 255 kg (324 656 lbm) and a thrust-to-weight ratio of 1.2. Two seconds after lift-off, the vehicle initiated a programmed roll from the launch pad azimuth of 115° to the required flight azimuth of 60°. This flight azimuth was attained at T + 20 seconds. At T + 15 seconds, the vehicle began a pre-programmed pitchover maneuver which lasted through booster engine cutoff. The Centaur inertial guidance system was functioning during this time, but steering commands were not admitted to the Atlas flight control system until after booster engine staging.

The pitch program used to command the vehicle during the booster flight was provided by the Centaur guidance system. This pitch program, one of a series selected on the basis of measured prelaunch upper-air soundings, was stored in the Centaur airborne computer. Booster engine gimbal angles for thrust vector control did not exceed 2.4° during the atmospheric ascent.

Vehicle acceleration during the boost phase was according to the mission plan. Centaur guidance issued the booster engine cutoff signal when the vehicle acceleration reached 5.74 g's (specification 5.79±0.113). Three seconds later at T + 155.2 seconds, the Atlas programmer issued the staging command to separate the booster engine section from the vehicle. Staging transients were small, and the maximum vehicle angular rate in pitch, yaw, or roll did not exceed 3.11 degrees per second. Vehicle steering by the inertial guidance system was initiated about 5 seconds following Atlas booster engine staging. At the start of guidance steering, the vehicle was slightly off the required steering vector by about 1° nose low in pitch and 1° nose right in yaw. The guidance system steered the vehicle to the proper vector and issued commands to continue the pitchover maneuver during the Atlas sustainer flight phase.

Insulation panels were jettisoned during the sustainer flight phase at T + 196.8 seconds. All four panels were severed by the shaped charges and fell away from the vehicle. Unlike previous Atlas-Centaur flights, the nose fairing was jettisoned after Centaur main engine start, rather than during Atlas phase of flight. Vehicle angular rates resulting from the jettisoning of the insulation panels were insignificant.

Sustainer and vernier engine system performance was satisfactory throughout the flight. Sustainer engine cutoff was initiated by liquid-oxygen depletion at T + 234.5 seconds. Maximum vehicle acceleration just prior to sustainer engine cutoff was 1.57 g's. Coincident with sustainer engine cutoff, the guidance steering commands to the Atlas flight control system were disabled, allowing the vehicle to coast in a free-flight mode. This guidance mode prevented gimbaling the Centaur main engines and allowed the engines to be centered to maintain clearance between the engines and the interstage adapter during Atlas-Centaur separation.

The Atlas staging command was issued by the flight programmer at T + 236.4 seconds. A shaped-charge firing cut the interstage adapter to separate the two stages. Eight retrorockets on the Atlas then fired to move the Atlas stage away from the Centaur. The transients during separation were small, and the maximum angular rate imparted to the vehicle did not exceed 0.5 degree per second.

CENTAUR FLIGHT PHASE

The main engine start sequence for the Centaur stage was initiated prior to sustainer engine cutoff. Propellant boost pumps were started at T + 199.2 seconds and allowed to come up to speed. To prevent boost pump cavitation during the near-zero-gravity period from sustainer engine cutoff until main engine start at T + 246.0 seconds, the required net positive suction pressure was provided by pressure pulsing the propellant tanks with helium. Eight seconds prior to main engine start, the Centaur programmer issued prestart commands for engine firing. Centaur main engines were gimballed to zero. Engine prestart valves were opened to flow liquid hydrogen through the lines to chill the engine turbopumps. Chilldown of the turbopumps ensured against cavitation during pump acceleration and made possible a uniform and rapid thrust build-up after engine ignition. At T + 246.0 seconds, the ignition command was issued by the flight programmer and engine thrust increased to full flight levels.

Guidance steering for the Centaur stage was enabled at T + 250 seconds. The total residual angular rates and disturbing torques induced during the Atlas-Centaur staging interval resulted in only a slight vehicle drift off the steering vector. This attitude drift error was corrected within 1 second after start of guidance steering.

The nose fairing was unlatched and jettisoned at T + 257.6 seconds. Disturbances due to separation caused a slight roll disturbance of 1.76 degrees per second; but this disturbance was damped out within 2 seconds. About 9 seconds after fairing jettison, the guidance system commanded a yaw maneuver (dog leg) to realine the velocity vector to meet the OAO-II orbit inclination requirement of 35°. Main engine shutdown was commanded by Centaur guidance at T + 698.2 seconds. Orbital insertion occurred approximately over Bermuda at an altitude of about 772 kilometers (417 n mi).

The Centaur engine burn time to establish the parking orbit was about 13 seconds longer than expected. The additional firing time was necessary to compensate for an apparent low thrust level. The propellant utilization system performed satisfactorily and accurately controlled the fuel and oxidizer flow rates to the engines.

SPACECRAFT SEPARATION

Coincident with the main engine cutoff, guidance steering commands were temporarily disabled, and the hydrogen peroxide vehicle rate control system was activated. Angular rates resulting from engine shutdown transients were small and were quickly damped to rates less than the control threshold of 0.2 degree per second. The residual vehicle motion below the rate threshold allowed only a negligible drift in vehicle attitude. This drift did not interfere with the subsequent spacecraft separation.

The Centaur - OAO-II coasted in a near-zero-gravity field for about 50 seconds. During this time, commands were given by the Centaur programmer to the spacecraft to deploy solar panels, to extend balance arms, and to arm the spacecraft separation pyrotechnics. All commands were properly received by the spacecraft.

At T + 748.3 seconds, the command for spacecraft separation was given. The hydrogen peroxide vehicle rate control system was commanded off, the pyrotechnically operated latches were fired, and the separation springs pushed the OAO-II away from the Centaur. The maximum angular rate of the Centaur was 0.37 degree per second just prior to, and 0.65 degree per second just following spacecraft separation. These rates are well within the maximum allowable rate of 1 degree per second for the Centaur at the time of spacecraft separation.

CENTAUR RETROMANEUVER

Following spacecraft separation, it was necessary for the Centaur stage to perform a reorientation and retrothrust maneuver in order to alter its orbit and place it beyond the view of the OAO-II star tracker. The vehicle began to reorient to a new vector which was approximately the negative geocentric radius vector at T + 1055 seconds. At T + 1149 seconds, with the Centaur alined to the new radius vector, two vernier engines were fired to provide 444.8 newtons (100 lbf) of thrust for 49 seconds. As these engines were commanded off, two 13.3-newton (3-lbf) thrust engines were commanded on and fired continuously for the next 350 seconds. The final part of the retromaneuver was then accomplished by opening the engine prestart valves to allow residual propellants in the tanks to "blowdown" through the main engines. The prestart valves remained open, allowing tank pressures to decrease to zero, and the vehicle continued in orbit in a nonstabilized flight mode.

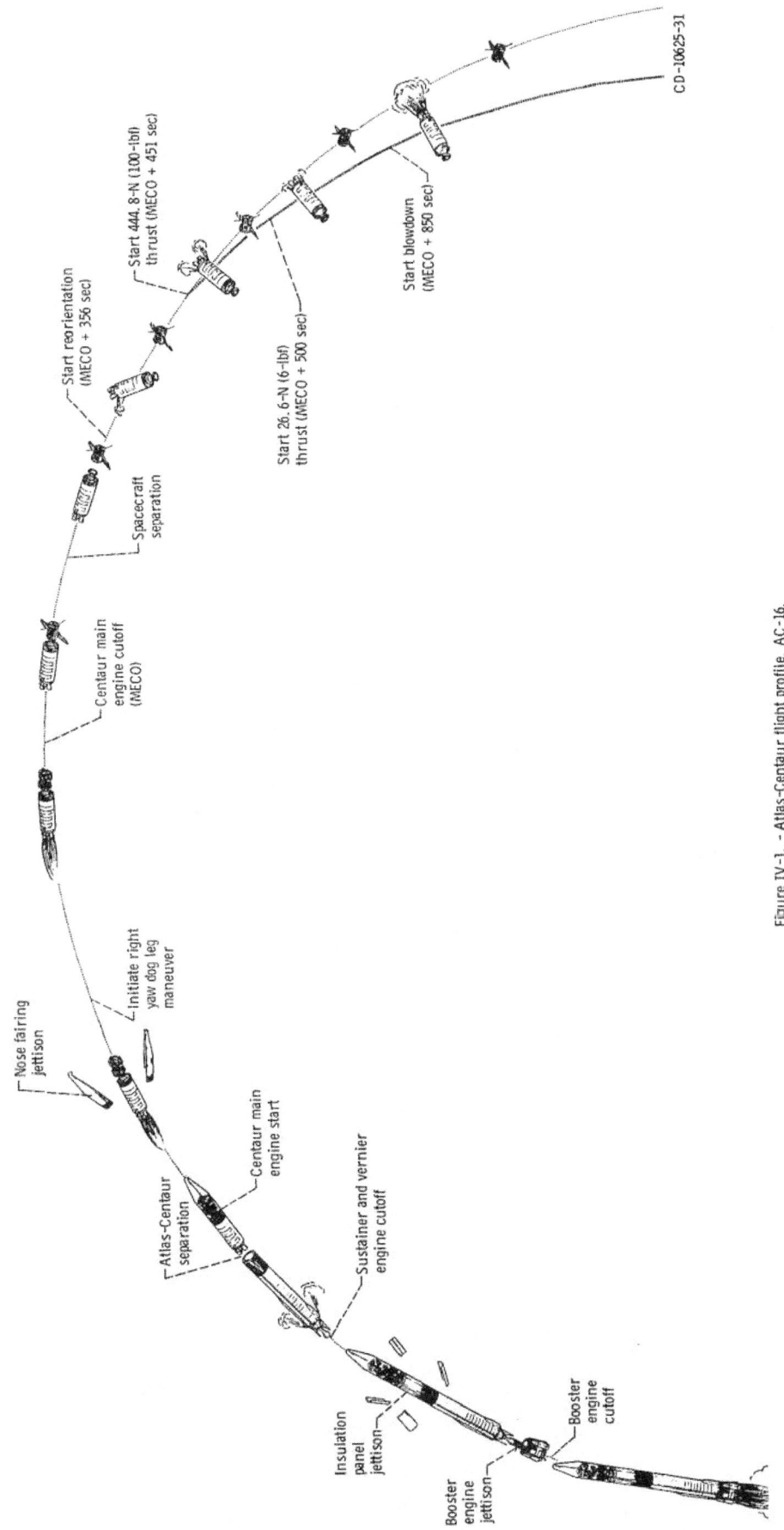

Start 444.8-N (100-lbf)
thrust (MECO + 451 sec)

Start reorientation
(MECO + 356 sec)

Start blowdown
(MECO + 850 sec)

Start 26.6-N (6-lbf)
thrust (MECO + 500 sec)

Spacecraft
separation

Centaur main
engine cutoff
(MECO)

Nose fairing
jettison

Initiate right
yaw dog leg
maneuver

Atlas-Centaur
separation

Centaur main
engine start

Sustainer and vernier
engine cutoff

Insulation
panel jettison

Booster
engine
jettison

Booster
engine cutoff

CD-10625-31

Figure IV-1. - Atlas-Centaur flight profile, AC-16.

17

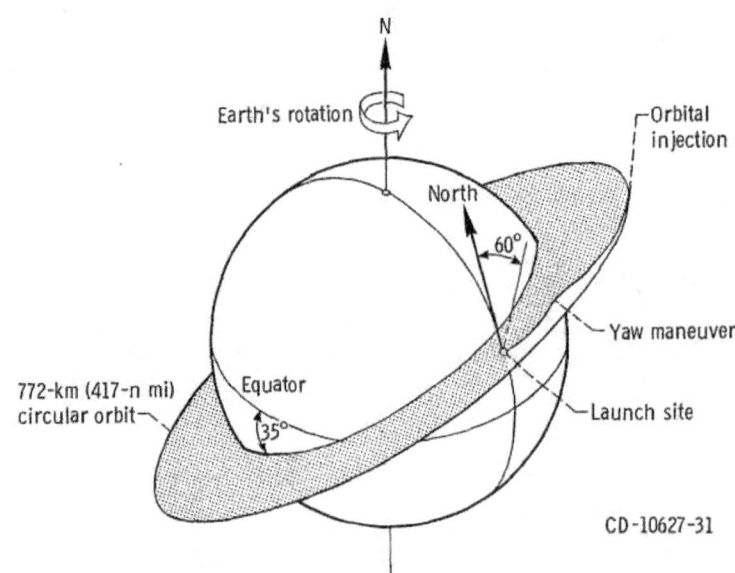

Figure IV-2. - OAO-II orbital trajectory, AC-16. Flight azimuth, 60°.

V. TRAJECTORY AND PERFORMANCE

by John J. Nieberding

MISSION PLAN

The mission plan for AC-16 was to launch the OAO-II spacecraft into a circular Earth orbit at an altitude, measured at the equator, of 772 kilometers (417 n mi) and at an orbital inclination to the equator of 35°. Achievement of this orbit required the steepest ascent trajectory ever flown by Atlas-Centaur. The ascent mode was direct ascent (i. e., Atlas and Centaur employed a nearly continuous powered phase, with the Centaur main engines firing only once). In order to achieve the required orbital inclination, a yaw maneuver around Bermuda was required during the Centaur firing period. Following main engine firing, the spacecraft was separated from Centaur to begin its postseparation sequence. The Centaur then performed a retromaneuver to meet a complex postseparation spacecraft viewing constraint.

TRAJECTORY RESULTS

Lift-off Through Atlas Booster Phase

Wind conditions at the launch site, based on data from a weather balloon sent aloft approximately 8 minutes after lift-off, were such that the ground wind was from the west/northwest (direction, 290°) at a speed of 11 kilometers per hour (6 knots). However, at an altitude of 12.2 kilometers (6.6 n mi), the speed had increased to a maximum of 102 kilometers per hour (89 knots) from 273°. These wind conditions required the use of pitch program 182 and yaw program 8 to minimize angle of attack through the booster phase of flight.

Radar tracking and Centaur guidance data indicated that the AC-16 flight path during the Atlas booster phase was very close to the predicted path. The transonic region, or the time span when the vehicle passes through Mach 1, occurred from about 60 to 65 seconds. During this period the axial load factor (thrust acceleration in g's) was nearly constant (fig. V-1). This is indicative of the vehicle's vibrations. Maximum dynamic pressure occurred at about T + 80 seconds (see fig. V-2). Atlas booster engine cutoff occurred at T + 152.1 seconds (see table V-I) when the vehicle axial load factor reached 5.74. (The axial load factor at booster engine cutoff is designed to be 5.7±0.113.) This cutoff was 0.8 second earlier than predicted.

At booster engine cutoff, the altitude was about 0.8 kilometer (0.4 n mi) lower than predicted, while the inertial velocity was about 96.6 kilometers per hour (88 ft/sec) lower than predicted (see figs. V-3 and V-4) (Note that fig. V-4 plots velocity relative to the atmosphere, not inertial velocity.) These deviations are well within tolerances.

Atlas Sustainer Phase

An abrupt decrease in acceleration occurred at T + 152.1 seconds when the booster engines cut off. This decrease is shown in figure V-1 and also in figure V-4, where a change in slope indicates a change in acceleration. A small but sudden increase in acceleration occurred at T + 155.2 seconds when the booster engine section, weighing 3378 kilograms (7448 lbm), was jettisoned. Following booster jettison, the constantly decreasing vehicle propellant weight caused the axial acceleration to increase smoothly until sustainer and vernier engine cutoff at T + 234.5 seconds, except for a small perturbation caused by jettisoning the 526-kilogram (1159-lbm) insulation panels at T + 196.8 seconds. This cutoff occurred at exactly the time predicted. At this time, the altitude was less than 0.30 kilometer (0.16 n mi) lower than expected; the inertial velocity was about 65.4 kilometers per hour (60 ft/sec) low. The axial acceleration at sustainer and vernier engine cutoff dropped abruptly to zero, indicating the loss of all thrust.

Centaur Main Engine Firing Phase

Atlas-Centaur separation was timed to occur 1.9 seconds after sustainer engine cutoff. Note in figure V-4 that the velocity decreased significantly between sustainer engine cutoff and Centaur main engine start. This velocity loss indicates a period of free-fall in a steep trajectory under the influence of gravity only. The increase in velocity and acceleration at Centaur main engine start (T + 246.0 sec) can be seen in figures V-4 and V-1, respectively.

After main engine start, the uniformly decreasing Centaur propellant weight caused the axial acceleration to increase smoothly until the 1115-kilogram (2458-lbm) split fairing - nose fairing combination was jettisoned at T + 257.6 seconds. The reason for jettisoning the nose fairing during the Centaur phase (as opposed to the usual jettison prior to sustainer engine cutoff) was twofold: (1) the acceleration level sensed at this time (about 0.75 g's) was nearer to the 1-g level at which the fairing had been tested than the level would have been during the late sustainer phase (about 1.5 g's), and (2) the probability of Atlas retrorocket exhaust impingement on the spacecraft optics was precluded by delaying the time of fairing jettison.

Shortly after the fairing was jettisoned, the Centaur main engines were gimbaled in order to perform a turn (yaw maneuver) to the right. This maneuver was required to achieve a final orbit inclination of 35°. Figure V-5 illustrates the instantaneous trace of the point at which Earth impact would occur if the Atlas-Centaur vehicle would lose all thrust or jettison a piece of hardware. These traces are called instantaneous impact point (IIP) traces. The figure shows the elliptical areas of impact for the sustainer section and nose fairing for three launch azimuths of 80°, 67°, and 60°. In order to achieve 35° inclination without performing any yaw maneuver, a launch azimuth of 67° or 113° would have been required. Tracking constraints precluded a launch azimuth of 113°. With an azimuth of 67°, there was a high probability that the nose fairing would impact inside the zone around Bermuda protected by Range Safety. This same zone prevented launching at azimuths between 60° and 80°. The 60° launch azimuth was chosen because the launch vehicle had better performance capability at 60° than at 80°. This azimuth then required a yaw to the right, once past Bermuda, to essentially "parallel" the 67° IIP trace and thereby achieve the proper orbital inclination.

During the remainder of the Centaur phase, the axial acceleration and relative velocity increased uniformly until Centaur main engine cutoff. During this period, however, the acceleration was approximately 3.2 percent lower than predicted because of low Centaur thrust. This effect can be seen in figure V-4, where the velocities are lower than expected throughout the Centaur phase until the proper cutoff value had been achieved, 13.0 seconds later than predicted. Based on a trajectory reconstructed from postflight Centaur guidance data, altitude at main engine cutoff differed from the predicted value by less than 0.18 kilometer (0.10 n mi), and velocities agreed to within 1.2 kilometers per hour (1.1 ft/sec). The spacecraft separated from Centaur 50.1 seconds later. Comparisons of the spacecraft orbit actually achieved with the orbit predicted are presented in table V-II. The difference between the predicted parameters and those based on the Guidance Reconstructed Trajectory (GRT) is an indication of the errors caused by guidance equation inaccuracies and dispersions in the Centaur main engine shutdown impulse. The difference between the GRT data and those based on the Best Estimate of Trajectory (BET) is primarily due to errors in guidance hardware and BET tracking data. The difference between Goddard Space Flight Center's tracking data and the BET is primarily caused by tracking and orbit determination errors. One point is to be noted: Figure V-3 shows the altitude at main engine cutoff (and, hence, also at spacecraft separation since the orbit is nearly circular) to be 779 kilometers (420 n mi). Yet table V-II quotes a perigee altitude of only about 772 kilometers (417 n mi). The reason is that the 779 kilometers (420 n mi) is referenced to an oblate Earth. At main engine cutoff, the actual altitude above the real, oblate Earth was 779 kilometers (420 n mi). However, by the time the vehicle crossed the equator, keeping a nearly constant radius, because the Earth has a greater diameter at the equator its altitude was about 772 kilometers (417 n mi). Thus, the apogee and perigee altitudes quoted in table V-II refer to the maximum and minimum altitudes respectively, when the vehicle passes over the equator.

Centaur Retromaneuver

Operation of the OAO-II spacecraft was restricted to periods when the Centaur tank is out of view of the spacecraft star sensor. Figure V-6 illustrates the pertinent geometry. When a line is drawn from the spacecraft 14° above the spacecraft horizon and then rotated about the spacecraft, maintaining the 14° angle, a conical volume is generated. If the Centaur tank, after separation from the spacecraft, is not within this volume, it is potentially in view of the spacecraft star sensor. A Centaur retromaneuver capable of assuring that the tank will be permanently out of view was not possible. The only possible maneuvers resulted in the Centaur's going out of view at some time after separation but again coming back into view at some later time. Analysis revealed that in order to minimize the probability of the Centaur staying in view for excessively long periods of time, the retromaneuver should be designed to yield a time to get out of view of 5.7 days after spacecraft separation with a corresponding return to view 64.3 days later. Centaur guidance data indicated that the Centaur went out of view in 6.5 days and returned into view 73.5 days later. Radar tracking data from the North American Air Defense Command (NORAD) showed a time to get out of view of 5.5 days, and a return to view 61.5 days later. Because of potentially large dispersions in the impulse supplied by the Centaur retromaneuver, these data are considered to be in good agreement with predicted values. (Predicted and actual Centaur postretromaneuver orbital parameters are given in table V-III.)

TABLE V-I. - FLIGHT EVENTS RECORD, AC-16

Event	Programmer time, sec	Preflight time, sec	Actual time, sec
Lift-off (2-in. (5.08-cm) motion)	T + 0	T + 0	T + 0
Start roll	T + 2.0	T + 2.0	T + 2.0
Start pitchover	T + 15.0	T + 15.0	T + 15.0
Stop roll	T + 20.0	T + 20.0	T + 20.0
Booster engine cutoff (BECO)	BECO	T + 152.9	T + 152.1
Jettison booster engines	BECO + 3.1	T + 156.0	T + 155.2
Admit guidance steering	BECO + 8.0	T + 160.9	T + 160.1
Jettison insulation panels	BECO + 45.0	T + 197.9	T + 196.8
Start Centaur booster pumps	BECO + 47.0	T + 199.9	T + 199.2
Sustainer engine cutoff (SECO); inhibit guidance steering	SECO	T + 234.6	T + 234.5
Atlas-Centaur separation	SECO + 1.9	T + 236.5	T + 236.4
Fire Atlas retrorockets	SECO + 2.0	T + 236.6	T + 236.5
Centaur prestart	SECO + 3.5	T + 238.1	T + 238.0
Centaur main engine start	SECO + 11.5	T + 246.1	T + 246.0
Readmit guidance steering	SECO + 15.5	T + 250.1	T + 250.0
Jettison nose fairing	SECO + 23.5	T + 258.1	T + 257.6
Centaur main engine cutoff (MECO)	MECO	T + 685.2	T + 698.2
Deploy solar paddles	MECO + 10.0	T + 695.2	T + 708.2
Extend balance booms	MECO + 25.0	T + 710.2	T + 723.2
Spacecraft separate	MECO + 50.0	T + 735.2	T + 748.3
Start reorientation to retrovector	MECO + 356.0	T + 1041.2	T + 1055.2
Start propellant settling engines	MECO + 451.0	T + 1136.2	T + 1149.2
Stop propellant settling engines; start propellant retention engines	MECO 500.0	T + 1185.2	T + 1198.2
Start discharge of Centaur residual propellants (blow-down)	MECO + 850.0	T + 1535.2	T + 1548.2

TABLE V-II. - OAO-II ORBIT PARAMETERS, AC-16

Parameter	Units	Predicted value	Actual value		
			GRT[a]	BET[b]	GSFC[c]
Epoch	sec from lift-off	736.0	748.3	748.3	748.3
Semimajor axis	km	7150.018	7150.885	7151.799	7151.979
	n mi	3860.701	3861.169	3861.663	3861.760
Inclination	deg	34.9982	34.9990	34.9811	34.9815
Eccentricity	----	0.000089	0.000083	0.00022	0.00030
Apogee altitude	km	772.586	773.210	775.197	775.950
	n mi	417.163	417.500	418.573	418.980
Perigee altitude	km	771.312	772.114	772.045	771.654
	n mi	416.475	416.908	416.871	416.660
Period	min	100.283	100.299	100.318	100.322

[a]Guidance Reconstructed Trajectory obtained from telemetered Centaur guidance data.

[b]Best Estimate of Trajectory obtained from Eastern Test Range radar tracking data.

[c]Goddard Space Flight Center data obtained from Manned Space Flight Net tracking data.

TABLE V-III. - CENTAUR POSTRETROMANEUVER

ORBITAL PARAMETERS, AC-16

Parameter	Units	Predicted value	Actual value
Epoch	sec from lift-off	5485.1	7842.0
Semimajor axis	km	7152.0	7148.3
	n mi	3861.8	3859.8
Inclination	deg	35.0293	34.9908
Eccentricity	----	0.00550	0.00645
Apogee altitude	km	813.4	814.5
	n mi	439.2	439.8
Perigee altitude	km	734.7	729.3
	n mi	396.7	393.8
Period	min	100.328	100.247

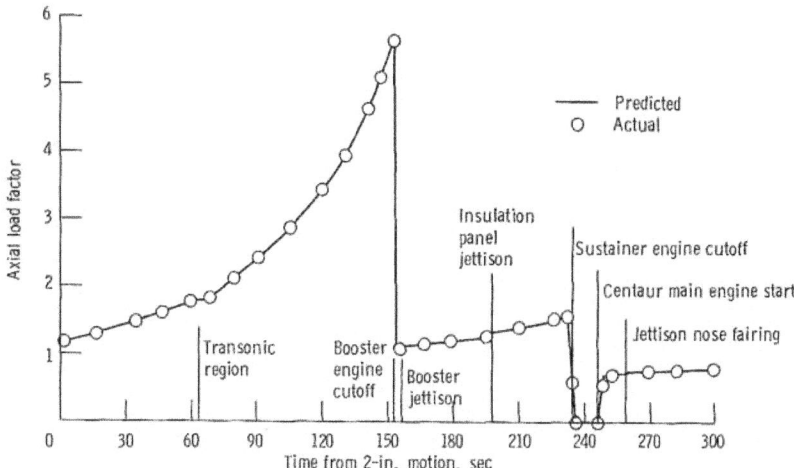

Figure V-1. – Axial load factor as function of time, AC-16.

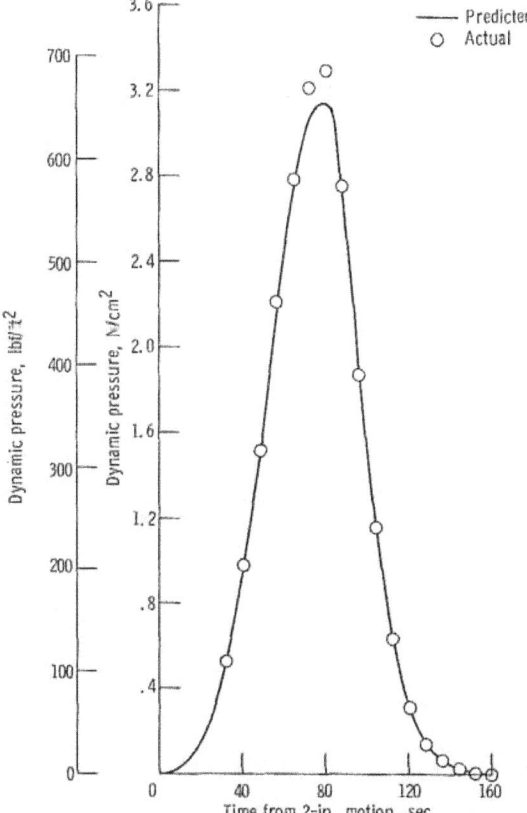

Figure V-2. – Dynamic pressure as function of time, AC-16.

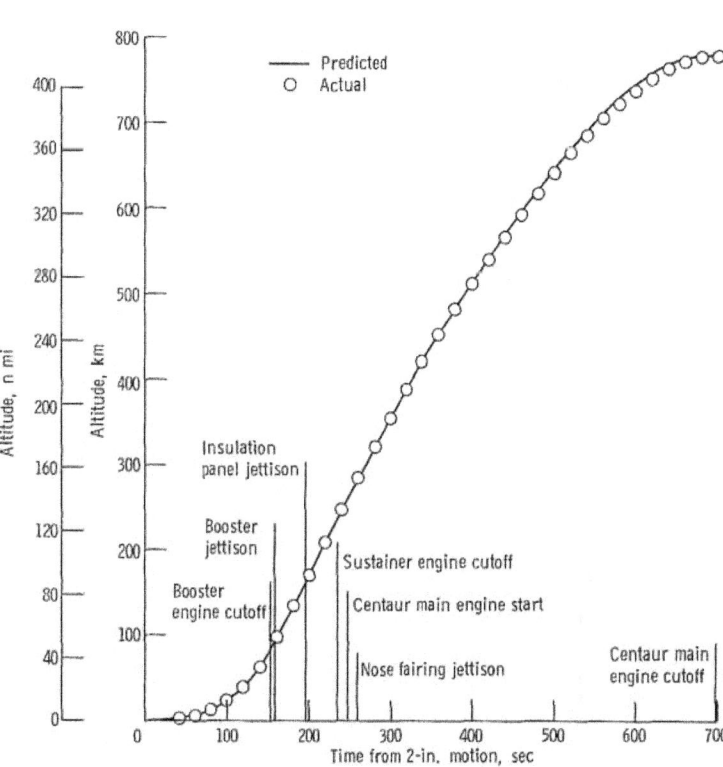

Figure V-3. – Altitude as function of time, AC-16.

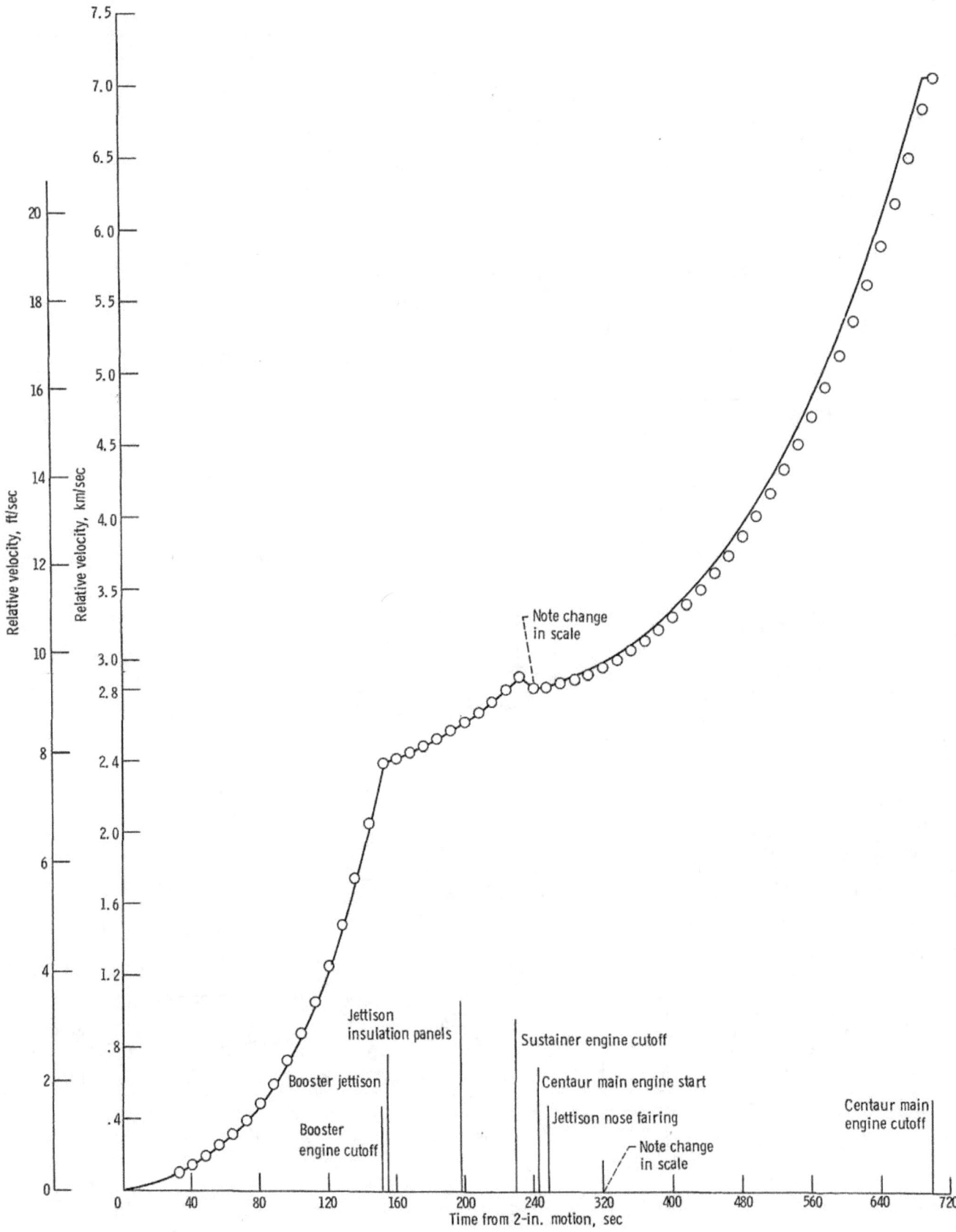

Figure V-4. - Relative velocity as function of time, AC-16.

26

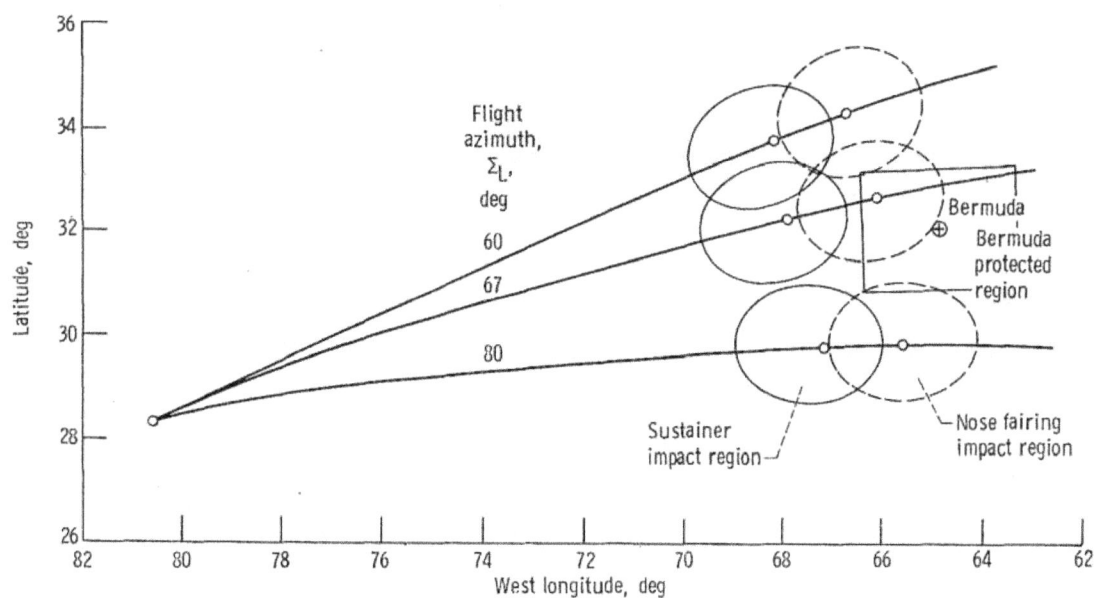

Figure V-5. – Instantaneous impact point traces for OAO-II mission.

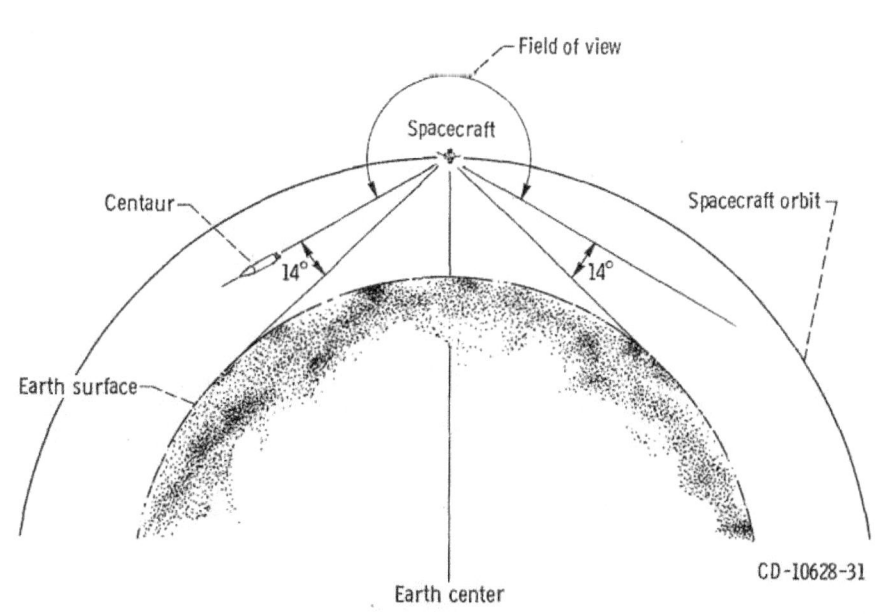

CD-10628-31

Figure V-6. – OAO-II field ov view.

VI. LAUNCH VEHICLE SYSTEM ANALYSIS

PROPULSION SYSTEMS

by Kenneth W. Baud, Charles H. Kerrigan,
Ronald W. Ruedele, and Donald B. Zelten

Atlas

System description. - The Atlas engine system (fig. VI-1) consists of a booster engine, a sustainer engine, two vernier engines, an engine start system (pressurization and auxiliary propellant), and an electrical control system. The engines are of the single-burn type. During engine start, electrically fired pyrotechnic igniters are used to ignite the gas generator propellants for driving the turbopumps, and hypergolic igniters are used to ignite the propellants in the thrust chambers of the booster, sustainer, and vernier engines. The propellants are liquid oxygen and RP-1 (kerosene).

The booster engine, rated at 1494×10^3 newtons (336×10^3 lbf) thrust at sea level, is made up of two gimbaled thrust chambers, propellant valves, two oxidizer and two fuel turbopumps driven by one gas generator, a lubricating oil system, and a heat exchanger. The sustainer engine, rated at 258×10^3 newtons (58×10^3 lbf) thrust at sea level, consists of a thrust chamber, propellant valves, one oxidizer and one fuel turbopump driven by a gas generator, and a lubricating oil system. The entire sustainer engine system gimbals. Each vernier engine is rated at 2.98×10^3 newtons (670 lbf) thrust at sea level, and propellants are supplied from the sustainer turbopump. The vernier engines gimbal for roll control.

The engine start system consists of two small propellant tanks (each ~51 cm (20 in.) in diam) and a pressurization system.

System performance. - The performance of the Atlas propulsion system for the OAO-II mission was satisfactory. During the engine start phase, valve opening times and starting sequence events were within tolerances. The flight performance of the engines was evaluated by comparing measured parameters with the expected values. The data are tabulated in table VI-I. Booster engine cutoff occurred at T + 152.1 seconds when the axial acceleration reached 5.74 g's. Sustainer engine cutoff and vernier engine cutoff occurred at T + 234.5 seconds and were due to liquid-oxygen depletion, the planned shutdown mode. Transients were normal during shutdown of all engines.

Centaur Main Engines

System description. - Two RL10A-3-3 engines (identified as C-1 and C-2) are used to provide thrust for the Centaur stage. Each engine has a thrust chamber which is regeneratively cooled and turbopump fed. Propellants are liquid oxygen and liquid hydrogen injected at an oxidizer-to-fuel mixture ratio of approximately 5 to 1. Engine rated thrust is 66 700 newtons (15 000 lbf) at an altitude of 61 000 meters (200 000 ft) and a design combustion chamber pressure of 274 newtons per square centimeter (400 psi). Thrust chamber nozzle expansion area ratio is 57 to 1, and design specific impulse is 442 seconds.

These engines use a "bootstrap" process: pumped fuel, after circulating through the thrust chamber tubes, is expanded through a turbine which drives the propellant pumps (see engine system schematic, fig. VI-2). This routing of fuel through the thrust chamber tubes serves the dual purpose of cooling the thrust chamber walls and of adding energy to the fuel prior to expansion through the turbine. After passing through the turbine, the fuel is injected into the combustion chamber. The pumped oxidizer is supplied directly to the combustion chamber after passing through the propellant utilization (mixture ratio control) valve.

The thrust level is maintained by regulating the amount of fuel bypassed around the turbine as a function of combustion chamber pressure. Ignition is accomplished by means of a spark igniter recessed in the propellant injector face. Starting and stopping are controlled by pneumatically operated valves. Helium pressure to these valves is supplied through engine mounted solenoid valves which are controlled by electrical signals from the vehicle control system.

System performance. - In-flight turbopump chilldown was commanded 8.0 seconds prior to main engine start. This chilldown was accomplished by opening both the oxidizer and the fuel pump inlet valves, allowing the propellants to flow into the engines. The oxidizer passed through the pump and into the combustion chamber; the fuel passed through the pump and overboard through two separate cooldown valves, one located downstream of each pump stage. This chilldown successfully prevented cavitation of the turbopumps during the start transient.

Main engine start was commanded at T + 246.0 seconds. The engine start transient was normal. Engine acceleration times to 90-percent chamber pressure were 1.48 and 1.41 seconds for the C-1 and C-2 engines, respectively. Total impulse from the main engine start command through 2.0 seconds of operation was calculated to be 43 300 and 45 900 newton-seconds (9740 and 10 360 lbf-sec) for the C-1 and the C-2 engines, respectively. The differential impulse between engines was well within the allowable of 25 000 newton-seconds (5700 lbf-sec).

The pump inlet pressures remained well above saturation at all times during engine operation. The pump inlet net positive suction pressures (i. e., total pressure minus saturation pressure) at different times throughout the engine firing period are presented in table VI-II. At no time did the levels drop below the minimum required of 2.76 and 5.52 newtons per square centimeter (4.0 and 8.0 psi) for the fuel and oxidizer pumps, respectively.

Steady-state operating conditions during the engine firing period are presented in table VI-III. The readings of chamber pressure taken while the propellant utilization valves were "nulled" (during the first 90 and last 13 seconds of main engine operation) were approximately 6.9 and 5.5 newtons per square centimeter (10 and 8 psi) less than those obtained on the final acceptance tests of the C-1 and C-2 engines, respectively. If these lower values of chamber pressure were indicative of a change in engine performance, other engine internal parameters should substantiate this change. However, the remaining engine internal parameters indicated normal levels: fuel venturi upstream pressures, fuel turbine inlet temperatures, and oxidizer pump speeds were fairly consistent with the values obtained during the final ground acceptance tests. These values do not substantiate the lower-than-expected values of chamber pressure, and the cause of the lower-than-expected chamber pressure is unknown.

Main engine performance in terms of thrust, specific impulse, and mixture ratio is presented in table VI-IV. Performance was calculated using the Pratt & Whitney characteristic velocity (C*) technique. An explanation of this technique is presented in appendix B of this report. The lower levels of thrust and higher levels of specific impulse, compared to acceptance test data, are the result of the low levels of chamber pressure previously discussed.

Main engine cutoff was commanded at T + 698.2 seconds, and the shutdown sequence was normal. Main engine cutoff total impulse was calculated to be 13 620 newton-seconds (3070 lbf-sec). This value compares favorably with the predicted value of 13 550±775 newton-seconds (3055±175 lbf-sec). The main engine firing duration of 452.2 seconds was 13.1 seconds longer than predicted. No explanation is available for the long firing duration; however, there was no adverse effect on launch vehicle performance.

Retromaneuver. - A vehicle retrothrust operation ("blowdown") was started 850 seconds following main engine cutoff. The retrothrust was provided by opening the pump inlet valves and allowing the propellants in the tanks to discharge through the main engine system continuously until the loss of vehicle power, which caused the inlet valves to close, terminating this operation. Engine pump inlet pressures and temperatures responded as expected to this operation. These parameters are presented for the beginning and end of the first orbital pass in table VI-V. At the end of the first orbital pass, both Centaur propellant tanks had been depleted of their liquid supply but were still discharging gases.

Centaur Boost Pumps

System description. - A single boost pump is used in each propellant tank to supply propellants to the main engine turbopumps at the required inlet pressures. Each boost pump is a mixed-flow centrifugal type and is powered by a hot-gas-driven turbine. The hot gas consists of superheated steam and oxygen from the catalytic decomposition of 90-percent-concentration hydrogen peroxide. Constant power is maintained on each turbine by metering the hydrogen peroxide through fixed orifices upstream of the catalyst bed. An overspeed speed control system is provided on each turbine. However, on this flight they were disconnected. The complete boost pump and hydrogen peroxide supply systems are shown in figures VI-3 to VI-6.

Boost pump performance. - Performance of the boost pumps was satisfactory for the entire flight. Boost pump start was initiated at T + 199.2 seconds. Boost pump operation was terminated simultaneously with main engine cutoff at T + 698.2 seconds.

The turbine inlet pressure delay time (time from boost pump start signal to time of first indication of turbine inlet pressure rise) was 1 second for both turbines. Steady-state turbine inlet absolute pressure for the oxidizer boost pump was 65.7 newtons per square centimeter (95.4 psi). The expected value based on prelaunch ground tests was 66.1 newtons per square centimeter (96.0 psi). The average steady-state turbine inlet absolute pressure for the fuel boost pump was 67.9 newtons per square centimeter (98.6 psi) with peak-to-peak pressure oscillations of 17.2 newtons per square centimeter (25 psi) superimposed. The expected value based on prelaunch test data was 70.3 newtons per square centimeter (102 psi). The pressure oscillations were noted during the prelaunch testing on this particular turbine, and have also been noted on several other turbines during both ground testing and flights. There was no detrimental effect on the turbine speed.

Fuel boost pump turbine speed is shown in figure VI-7. The steady-state speed was approximately 1300 rpm higher than the prelaunch test value of 39 900 rpm. Oxidizer boost pump turbine speed is shown in figure VI-8. The steady-state speed was approximately 1500 rpm higher than the prelaunch test value of 32 900 rpm. Boost pump turbine speeds have been consistently higher than the acceptance test values on virtually all previous Centaur flights. These differences are due to the inability to accurately simulate the flight conditions during prelaunch ground tests.

Turbine bearing temperatures for the fuel and oxidizer boost pumps are shown in figures VI-9 and VI-10, respectively. The maximum values were comparable to values recorded on previous single-burn flights.

Liquid-oxygen temperature at the boost pump inlet is shown in figure VI-11. The temperature rise between boost pump start and main engine start was a result of warm fluid being returned to the sump area through small bleed holes in the pump volute casting. The temperature dropped shortly after main engine start, when the flow to the main engines began and the warm liquid was removed from the sump. Since there was no pressurization of the tank after main engine start, the tank pressure decreased as the liquid oxygen was drained from the tank. There was a corresponding decrease in the liquid-oxygen bulk temperature as liquid continually evaporated to maintain the liquid saturated at the ullage pressure.

Liquid-hydrogen temperature at the boost pump inlet is shown in figure VI-12. The rise in temperature between T - 0.5 minutes and T + 1.5 minutes was caused by lockup of the hydrogen tank primary (lower range) vent valve. Heat input to the liquid during this time resulted in a tank pressure increase and a corresponding bulk temperature increase. When the vent valve was unlocked, the liquid hydrogen resaturated at the primary vent valve pressure setting, and a sharp drop in temperature resulted. The continued decrease in temperature after main engine start was a result of propellant outflow. As the pressure in the tank decreased, the liquid hydrogen resaturated at the reduced pressures.

Hydrogen Peroxide Engine and Supply System

System description. - The hydrogen peroxide system (figs. VI-4 and VI-13) consists of 14 thruster engines, a supply bottle, and interconnecting tubing to the engines and boost pump turbines. The engines are used after main engine cutoff. Four 222.4-newton- (50-lbf-) thrust engines and four 13.3-newton- (3-lbf-) thrust engines are used primarily for propellant settling and retention and for retromaneuver. Two clusters, each of which consists of two 15.6-newton- (3.5-lbf-) thrust engines and one 26.7-newton- (6-lbf-) thrust engine, are used for attitude control (see table VI-XIII, GUIDANCE AND FLIGHT CONTROL SYSTEMS, for mode of operation). Propellant is supplied to the engines from a positive-expulsion, bladder-type storage tank which is pressurized with helium to an absolute pressure of about 210 newtons per square centimeter (305 psi) by the pneumatic system. The hydrogen peroxide is decomposed in the engine catalyst beds, and the hot decomposition products are expanded through converging-diverging nozzles to provide thrust. Hydrogen peroxide is also provided to drive the boost pump turbines. All the hydrogen peroxide supply lines, except for one short line, are equipped with heaters. However, on AC-16 and all other single-burn vehicles, the heaters are not required on the boost pump feedlines, and they are electrically disconnected.

Configuration changes were made on AC-16 as a result of suspected cryogenic leakage during the AC-17 flight. All of the openings in the liquid-oxygen-tank radiation shield were covered in the vicinity of the hydrogen peroxide bottle to shield the hydrogen peroxide system from any leakage from the liquid-oxygen tank. A fiber glass shield was also installed on the hydrogen-peroxide-bottle support yoke to protect the system from possible leakage from the liquid-oxygen-tank sump flanges and the liquid-oxygen supply ducting. Instrumentation was added to assist in the detection of any cryogenic leakage and to obtain temperature data for the boost pump hydrogen peroxide supply lines.

System performance. - The supply bottle was tanked with 106.5 kilograms (234.9 lbm) of hydrogen peroxide. The bottle absolute pressure at lift-off was 224 newtons per square centimeter (325 psi). The pressure decreased to about 214 newtons per square centimeter (310 psi) by T + 180 seconds and then remained constant throughout the flight. The decrease in pressure is normal because the pneumatic regulator is referenced to ambient pressure, which becomes essentially zero when the vehicle leaves the atmosphere.

Four of the engine chamber surfaces were instrumented for temperature measurement (fig. VI-14): two 15.6-newton- (3.5-lbf-) thrust engines (A-1 and A-2), one 13.3-newton- (3-lbf-) thrust engine (S-1), and one 222.4-newton- (50-lbf-) thrust engine (V-1). All engine temperature data indicated normal performance. Temperature changes verified engine firing as programmed and as required to maintain vehicle control. The data showed that at T + 1492 seconds, during the S-half-on mode of firing (see table VI-XIII, GUIDANCE AND FLIGHT CONTROL SYSTEMS), a large disturbing torque on the vehicle exceeded the control capability of the S-engines. A rise in the V-1 engine chamber temperature at this time verified that the 222.4-newton- (50-lbf-) thrust engine fired to maintain vehicle control. The disturbing torque was caused by venting of the liquid-hydrogen tank. On previous flights, the vent stack was jettisoned as a part of the nose fairing, and venting was accomplished through balanced thrust vent ducting. However, on AC-16, because of the configuration of the nose fairing, the stack was not jettisoned, and this disturbing torque was expected when venting occurred.

The A-1 and A-2 engine chamber temperatures, during the second orbital pass (T + 7370 to T + 8230 sec over the Canary telemetry station) showed that these engines were still firing and maintaining vehicle control.

Boost pump hydrogen peroxide supply line temperature evaluation. - Several transducers were installed on AC-16 to determine in-flight temperatures of the hydrogen peroxide supply lines for the boost pumps. Location and identification of the transducers are shown in figure VI-14. The temperature data obtained from the AC-16 flight are shown in figure VI-15.

In general, the supply line temperature data indicated no abnormal conditions throughout the flight. At T - 180 seconds, the temperature measurement CP346T, between the bottle and boost pump feed valve, showed a sharp temperature rise. This rise was a result of warm hydrogen peroxide being forced from the bottle and into the line at the time of bottle pressurization. Immediately after lift-off, all temperatures decreased due to termination of the warm gas conditioning supply to the Centaur thrust section. Venting of the conditioning gas from the thrust section during the boost phase also contributed to the temperature decreases after lift-off.

The radiation shield temperature measurement (CP350T) started to rise at T + 100 seconds due to thermal radiation from the Atlas-Centaur interstage adapter. The interstage adapter temperature increased during the boost phase as a result of aerodynamic heating. An increase in the radiation shield temperature rise rate occurred at boost pump start (T + 199.2 sec) because the liquid-oxygen boost pump turbine exhaust discharges into the enclosed Centaur thrust section. The radiation shield temperature then decreased abruptly at T + 236.4 seconds, when Atlas-Centaur separation occurred. A steady decrease in the radiation shield temperature then occurred until T + 1052 seconds. At this time, the temperature began to rise again, reflecting an increase in the solar radiation on the aft end of the Centaur when the reorientation maneuver was started. The radiation shield temperature appeared to reach a maximum at approximately T + 2100 seconds, and then began a slow decrease which continued through T + 8230 seconds.

At boost pump start, all hydrogen peroxide supply line temperatures rose abruptly as the warm peroxide from the supply bottle flowed through the lines. During boost pump operation, all line temperatures, except CP352T, stabilized at values near the temperature of the hydrogen peroxide in the bottle, which was 302 K (544° R). However, CP352T, which was located less than 5 centimeters (2 in.) from the face of the hot turbine, continued to rise slowly throughout the boost pump operating period, and was 321 K (578° R) at main engine and boost pump cutoff.

In the space environment, the unheated lines remote from the turbines cool due to radiation. The lines near the turbines are heated by radiation and conduction from the hot turbine housing. Temperature measurements (CP344T, CP345T, CP346T, CP347T, and CP351T) on the lines remote from the turbines showed a gradual cooling trend after main engine cutoff, as was expected. The temperature measurements (CP348T, CP349T, and CP352T) which were located near the hot turbines showed a definite warming trend after main engine cutoff, as was expected. Shortly after main engine and boost pump cutoff, essentially all of the supply line temperatures began to exhibit rather abrupt changes which continued throughout the coast until loss of telemetry data at T + 8230 seconds. These abrupt changes were caused by "slugging" of residual hydrogen peroxide remaining in the lines downstream of the feed valve after boost pump cutoff. As small pockets or "slugs" of residual hydrogen peroxide moved past a transducer on a relatively cool line, a temperature increase occurred. Conversely, as small pockets or "slugs" of residual hydrogen peroxide moved past a transducer on a relatively hot line, a temperature decrease occurred.

A long time period was required to empty the residual hydrogen peroxide from the lines because of the combined effect of the low-gravity environment and the small flow-restricting orifices located near the turbines (see fig. VI-4). At T + 6800 seconds, measurement CP348T began to rise sharply, indicating that the random outflow (slugging) of residual hydrogen peroxide through the hydrogen boost pump supply line had ceased and that the short section of line between the speed limiting valve and catalyst bed was empty. The temperature at this location increased to a maximum value of 373 K (670° R), and then began to decrease slowly as the turbine cooled. The "slugging" was still evident in the liquid-oxygen boost pump turbine supply lines until loss of data at T + 8230 seconds. However, the gradual warming trend shown by measurement CP352T from T + 7600 to T + 8230 seconds indicated that the liquid-oxygen boost pump turbine supply lines were also nearly empty.

A simultaneous change in several of the supply line temperatures occurred at T + 1492 seconds. This change was the result of residual hydrogen peroxide movement in the lines when a large vehicle disturbance occurred. Pressure rise in the hydrogen tank caused the secondary (upper range) vent valve to relieve, and gaseous hydrogen was vented overboard at T + 1492 seconds. The venting imparted a large disturbing torque on the vehicle.

Based on the hydrogen peroxide supply line temperature data, it was concluded that there were no cryogenic leaks near the hydrogen peroxide system on AC-16. It was also concluded that small quantities of residual hydrogen peroxide were retained in the boost pump supply lines (downstream of the feed valve) until loss of telemetry data at T + 8230 seconds. The liquid-hydrogen boost pump supply line between the speed limiting valve and catalyst bed became dry at T + 6800 seconds, resulting in a maximum line temperature of 373 K (670° R) at measurement CP348T.

Comparison of AC-16 and AC-17 temperature data. - Two of the AC-16 temperature measurements (CP344T and CP345T) were of particular interest because of a flight failure of AC-17. The boost pumps failed to operate for the second burn of the AC-17 flight; this failure was attributed to blockage of the hydrogen peroxide flow to the boost pump turbines. Cryogenic leakage and subsequent freezing of the hydrogen peroxide within the boost pump turbine supply lines was a primary suspect. Only two temperature measurements were installed on the hydrogen peroxide supply lines to the turbines for AC-17. These two measurements were CP344T and CP345T (also installed on AC-16).

Comparison of the temperature data from measurements CP344T and CP345T for both AC-16 and AC-17 revealed the same abrupt temperature changes after main engine cutoff on both flights. Two significant differences were noted: (1) on AC-17, CP344T showed a sharp temperature rise for 76 seconds after main engine cutoff that was not evident on AC-16, and (2) both CP344T and CP345T reached much lower temperatures on AC-16 than on AC-17. On AC-17, both CP344T and CP345T stayed within a temperature band between 302 and 316 K (545° and 570° R) during the entire coast period of 61 minutes. On AC-16, these same two measurements decreased steadily to values of 242 and 255 K (436° and 460° R) after 61 minutes of coast. These two differences noted in the temperature data were due to configuration differences between the two vehicles. On AC-17, 222. 4-newton- (50-lbf-) thrust engines were fired for 76 seconds immediately after main engine cutoff to settle propellants; on AC-16, these engines were not pro- grammed to fire during this time period. Convective heating due to impingement of the engine exhaust plumes on the supply line caused the temperature to rise on AC-17. Measurements CP344T and CP345T showed lower temperatures on AC-16 because the boost pump turbine supply line heaters were electrically connected on AC-17, but were disconnected on AC-16.

TABLE VI-I. - ATLAS PROPULSION SYSTEM PERFORMANCE, AC-16

Performance parameter	Unit	Expected operating range	Flight values at -		
			T + 10 sec	Booster engine cutoff, T + 152.1 sec	Sustainer and vernier engine cutoff, T + 234.5 sec
Booster engine:					
Thrust chamber 1:					
Pressure, absolute	N/cm^2	386 to 410	398	401	(a)
	psi	560 to 595	577	581	(a)
Turbopump speed	rpm	6225 to 6405	6369	6393	(a)
Thrust chamber 2:					
Pressure, absolute	N/cm^2	386 to 410	400	403	(a)
	psi	560 to 595	580	584	(a)
Turbopump speed	rpm	6165 to 6345	6319	6334	(a)
Gas generator chamber	N/cm^2	351 to 382	373	373	(a)
pressure, absolute	psi	510 to 555	540	540	(a)
Sustainer engine:					
Thrust chamber pressure,	N/cm^2	469 to 493	486	476	476
absolute	psi	680 to 715	705	690	690
Gas generator	N/cm^2	407 to 473	441	441	441
discharge pressure,	psi	620 to 680	640	640	640
absolute					
Engine turbopump speed	rpm	10 025 to 10 445	10 287	10 196	10 301
Vernier engine thrust					
chamber absolute					
pressure:					
Engine 1	N/cm^2	172 to 183	181	178	181
	psi	250 to 265	262	258	262
Engine 2	N/cm^2	172 to 183	181	178	182
	psi	250 to 265	262	258	264

[a]Not applicable.

TABLE VI-II. - CENTAUR MAIN ENGINE PUMP INLET

NET POSITIVE SUCTION HEAD, AC-16

Time from main engine start, sec	Location							
	C-1 fuel pump		C-2 fuel pump		C-1 oxidizer pump		C-2 oxidizer pump	
	Net positive suction head							
	N/cm^2	psi	N/cm^2	psi	N/cm^2	psi	N/cm^2	psi
0	12.9	18.7	12.4	18.0	55.7	80.7	55.4	80.4
[a]~1.2	6.4	9.2	5.7	8.2	27.5	39.9	33.2	48.1
10	5.4	7.9	6.4	9.3	19.1	27.7	19.7	28.6
[b]90	7.1	10.3	6.6	9.5	19.9	28.8	20.4	29.6
[c]100	7.8	11.1	7.3	10.6	18.3	26.5	18.7	27.1
[d]112	6.9	10.0	5.5	7.9	21.4	31.0	21.9	31.7
[d]353	8.3	12.1	7.9	11.5	20.0	29.0	20.2	29.3
[b]450	5.6	8.2	6.7	9.9	20.5	29.7	20.7	30.0

[a]Minimum dip during start transient.

[b]This time was selected as being representative while the propellant utilization valves were nulled.

[c]This time was selected as being representative of when the propellant utilization valves were commanded to the oxidizer rich stop.

[d]This time was selected as representing the minimum and the maximum mixture ratio conditions the engines experienced during the portion of flight that the propellant utilization valves were controlling.

TABLE VI-III. – CENTAUR MAIN ENGINE OPERATING DATA, AC-16

Parameter	Units	Expected value	Time from main engine start[a], sec									
			90 C-1	90 C-2	100 C-1	100 C-2	112 C-1	112 C-2	353 C-1	353 C-2	450 C-1	450 C-2
Fuel pump inlet total pressure, absolute	N/cm^2	16.2 to 24.1	20.6	20.7	21.2	21.5	20.2	19.4	19.1	19.2	16.2	16.7
	psi	23.5 to 33.9	29.9	30.1	30.8	31.2	29.4	28.2	27.7	27.9	23.5	24.2
Fuel pump inlet temperature	K	20.1 to 21.6	21.2	21.5	21.3	21.5	21.2	21.4	20.4	20.8	20.4	20.2
	°R	36.1 to 38.8	38.2	38.6	38.3	38.6	38.2	38.5	36.8	37.4	36.7	36.4
Oxidizer pump inlet total pressure, absolute	N/cm^2	31.8 to 47.3	41.8	42.5	40.3	40.6	43.3	43.9	39.7	40.4	38.8	39.0
	psi	46.2 to 68.7	60.8	61.6	58.5	59.0	62.9	63.6	57.6	58.6	56.3	56.6
Oxidizer pump inlet temperature	K	95.3 to 98.3	98.1	98.1	98.1	98.1	98.1	98.0	97.0	97.0	96.1	96.1
	°R	171.5 to 176.9	176.6	176.6	176.6	176.6	176.6	176.5	174.8	174.8	173.0	173.0
Oxidizer pump speed	rpm	b12 163 ± 347	12 100.0	12 380.0	11 920.0	12 100.0	12 260.0	12 620.0	11 840.0	12 100.0	12 040.0	12 300.0
Fuel venturi upstream pressure, absolute	N/cm^2	b508 ± 17	505.0	515.0	494.0	502.0	515.0	540.0	491.0	508.0	505.0	516.0
	psi	b737 ± 25	732.0	748.0	716.0	729.0	749.0	784.0	714.0	737.0	734.0	750.0
Fuel turbine inlet temperature	K	b207 ± 12	208.0	220.0	228.0	238.0	185.0	200.0	221.0	231.0	204.0	215.0
	°R	b372 ± 22	374.0	394.0	410.0	428.0	332.0	359.0	398.0	416.0	368.0	388.0
Oxidizer injector differential pressure	N/cm^2	b31.7 ± 6.9	29.4	30.4	32.3	35.0	26.2	27.7	30.9	32.5	29.3	30.2
	psi	b46 ± 10	42.7	44.1	46.9	50.8	38.2	40.2	44.8	47.1	42.5	43.7
Engine chamber pressure, absolute	N/cm^2	b270.8 ± 3.7	263.0	263.0	264.0	265.0	261.0	263.0	263.0	265.0	263.0	263.0
	psi	b392.4 ± 5.4	381.0	382.0	383.0	384.0	379.0	382.0	382.0	384.0	381.0	382.0

a See footnotes to table VI-II for explanation of time sample selection.

b Expected value with nominal inlet conditions and nulled propellant utilization valve angle.

TABLE VI-IV. – CENTAUR MAIN ENGINE PERFORMANCE, AC-16

Performance parameter	Units	Acceptance value		Time from main engine start[a], sec									
		C-1	C-2	90 C-1	90 C-2	100 C-1	100 C-2	112 C-1	112 C-2	353 C-1	353 C-2	450 C-1	450 C-2
Thrust	N	66 400.0	66 500.0	64 690.0	65 160.0	66 320.0	66 320.0	63 720.0	64 380.0	65 570.0	66 040.0	64 560.0	65 010.0
	lbf	14 930.0	14 960.0	14 540.0	14 650.0	14 910.0	14 910.0	14 330.0	14 470.0	14 740.0	14 850.0	14 510.0	14 610.0
Specific impulse	sec	444.8	445.5	445.2	445.7	443.0	443.7	447.9	447.8	443.7	444.7	445.3	445.8
Mixture ratio	---	5.032	5.018	4.865	4.925	5.420	5.460	4.292	4.293	5.303	5.259	4.789	4.833

a See footnotes to table VI-II for explanation of time sample selection.

40

TABLE VI-V. - CENTAUR MAIN ENGINE PUMP INLET PRESSURES AND

TEMPERATURES DURING RETROTHRUST, AC-16

Parameter	Units	Time from start of retrothrust, sec							
		-5		5		300		6480	
		C-1	C-2	C-1	C-2	C-1	C-2	C-1	C-2
Fuel pump inlet pressure, absolute	N/cm^2	0	0	19.3	18.7	11.2	10.7	1.0	1.0
	psi	0	0	27.9	27.1	16.3	15.5	1.6	1.6
Fuel pump inlet temperature	K	>24.7	>24.8	21.7	21.8	>24.7	>24.8	>24.7	>24.8
	^{o}R	>44.4	>44.6	39.1	39.5	>44.4	>44.6	>44.4	>44.6
Oxidizer pump inlet pressure, absolute	N/cm^2	0	0	21.4	21.2	20.8	20.6	1.9	2.4
	psi	0	0	31.0	30.7	30.2	29.9	2.7	3.4
Oxidizer pump inlet temperature	K	97.0	97.2	96.9	97.0	97.0	97.0	>102.0	>102.0
	^{o}R	174.8	175.0	174.6	174.8	174.8	174.8	>183.3	>183.3

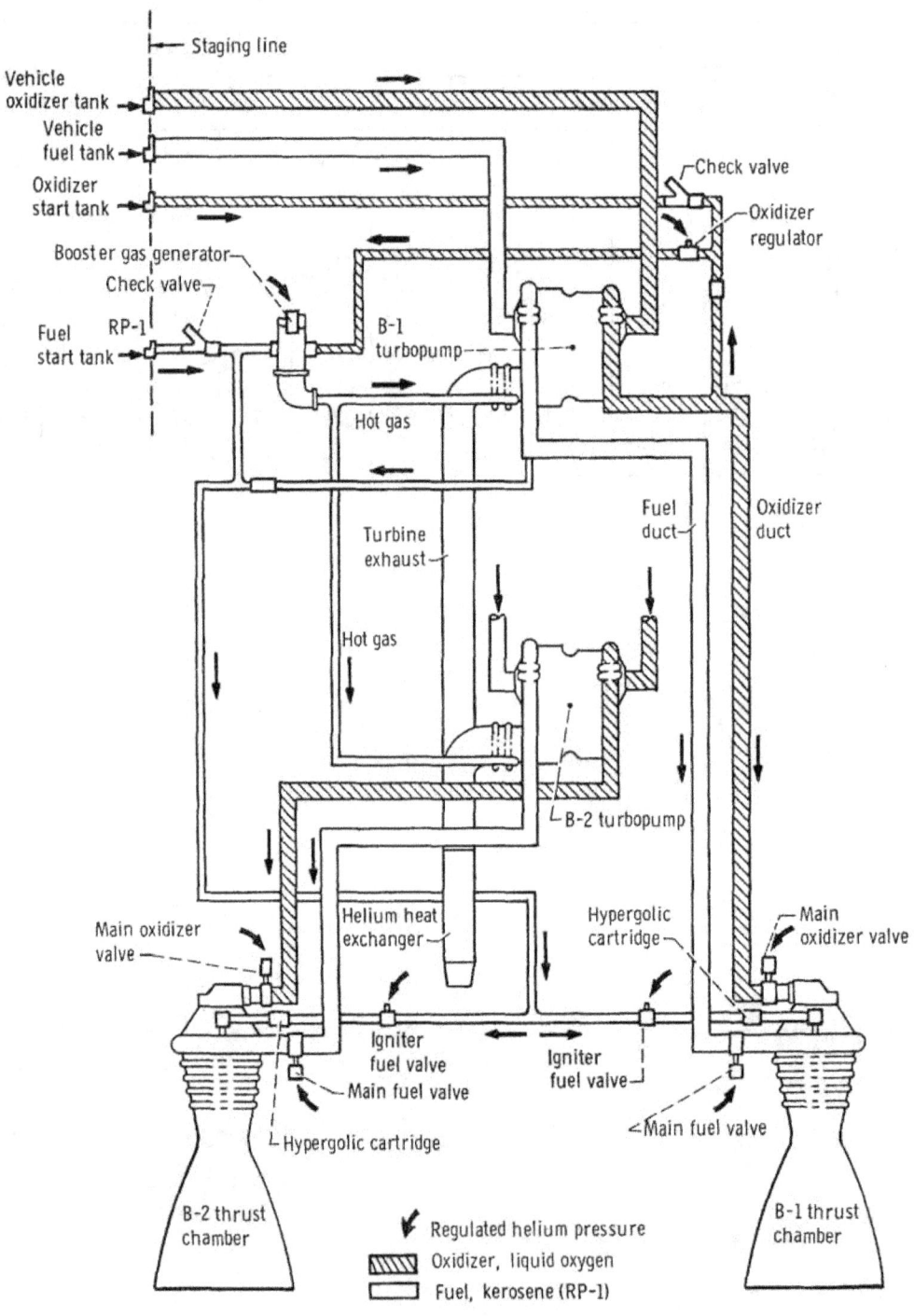

Staging line

Vehicle oxidizer tank

Vehicle fuel tank

Oxidizer start tank

Booster gas generator

Check valve

Fuel start tank — RP-1

Check valve

Oxidizer regulator

B-1 turbopump

Hot gas

Turbine exhaust

Hot gas

Fuel duct

Oxidizer duct

B-2 turbopump

Main oxidizer valve

Helium heat exchanger

Hypergolic cartridge

Main oxidizer valve

Igniter fuel valve

Main fuel valve

Igniter fuel valve

Main fuel valve

Hypergolic cartridge

B-2 thrust chamber

Regulated helium pressure

Oxidizer, liquid oxygen

Fuel, kerosene (RP-1)

B-1 thrust chamber

(a) Atlas vehicle booster engine.

Figure VI-1. - Atlas propulsion system, AC-16.

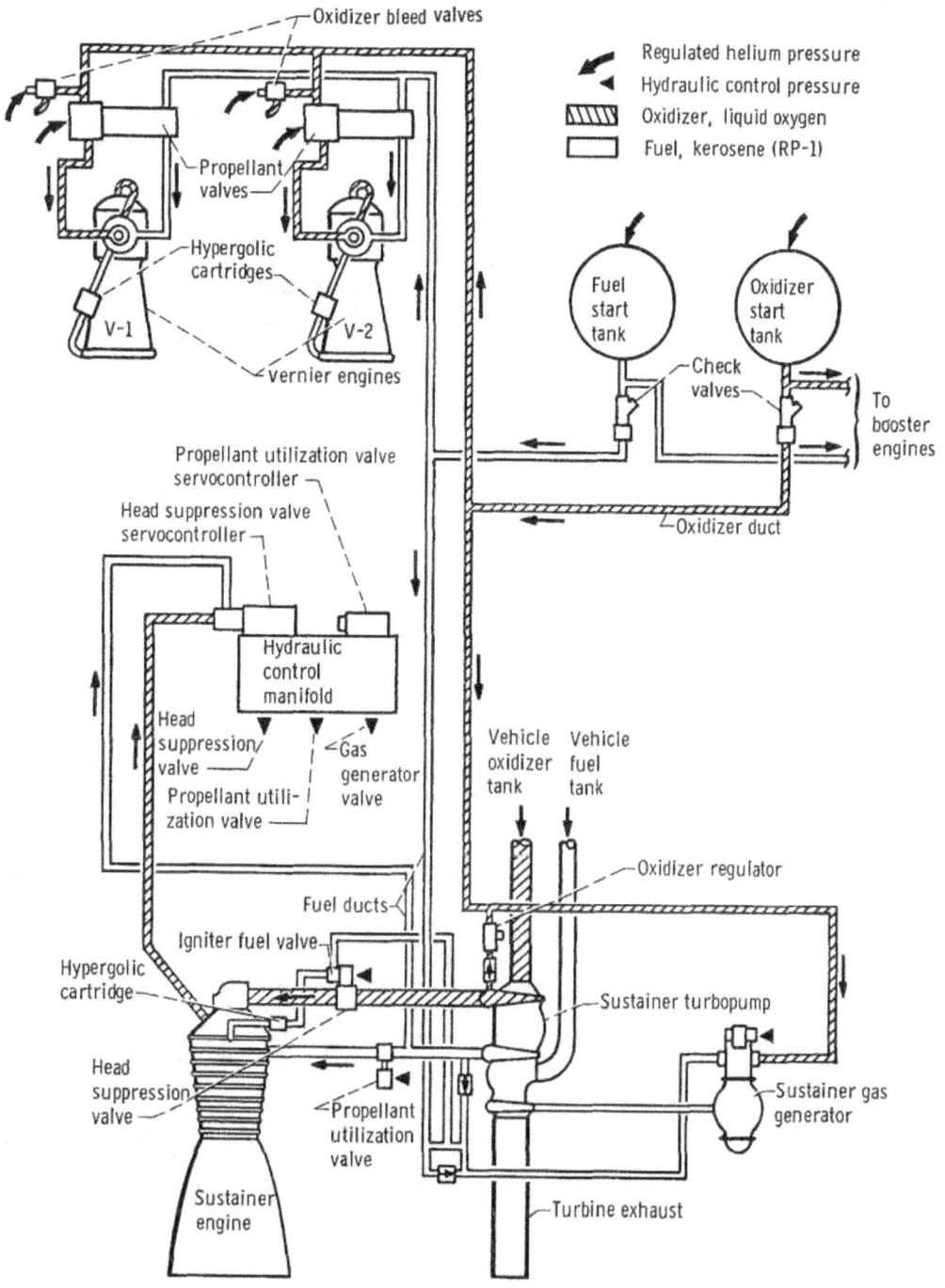

(b) Atlas vehicle sustainer and vernier engines.

Figure VI-1. - Concluded.

43

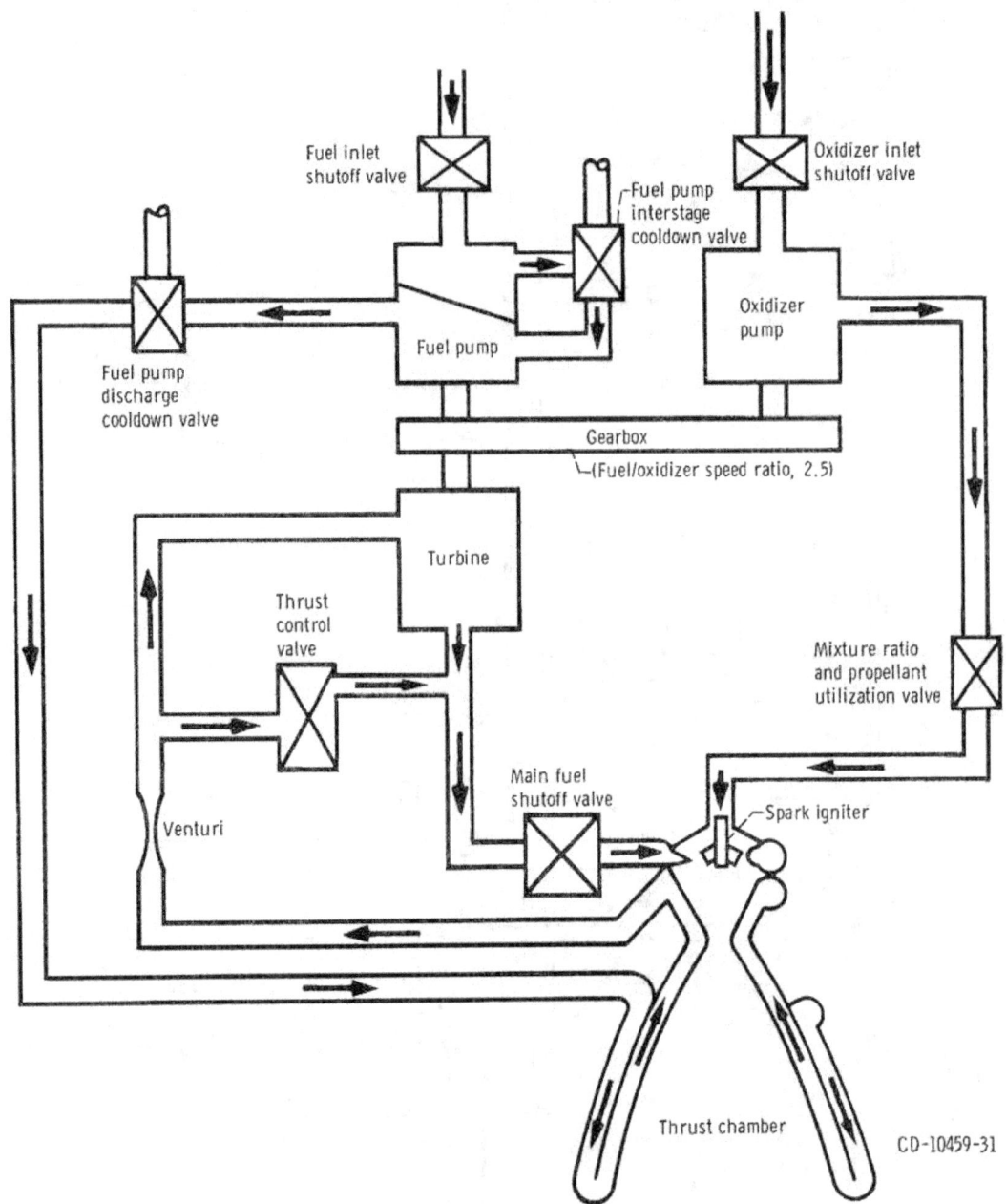

Figure VI-2. - Centaur propulsion system, AC-16.

CD-10459-31

44

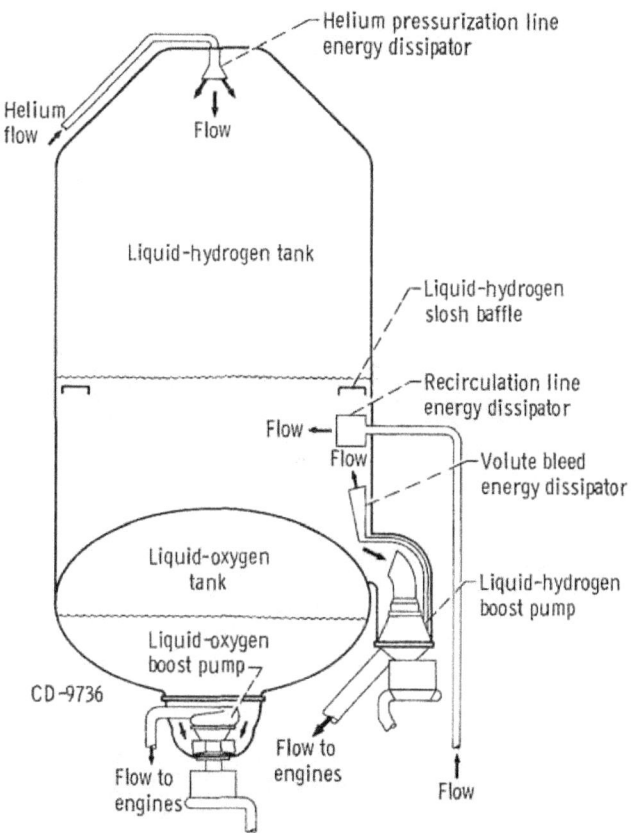

Figure VI-3. - Location of Centaur liquid-hydrogen
and liquid-oxygen boost pumps, AC-16.

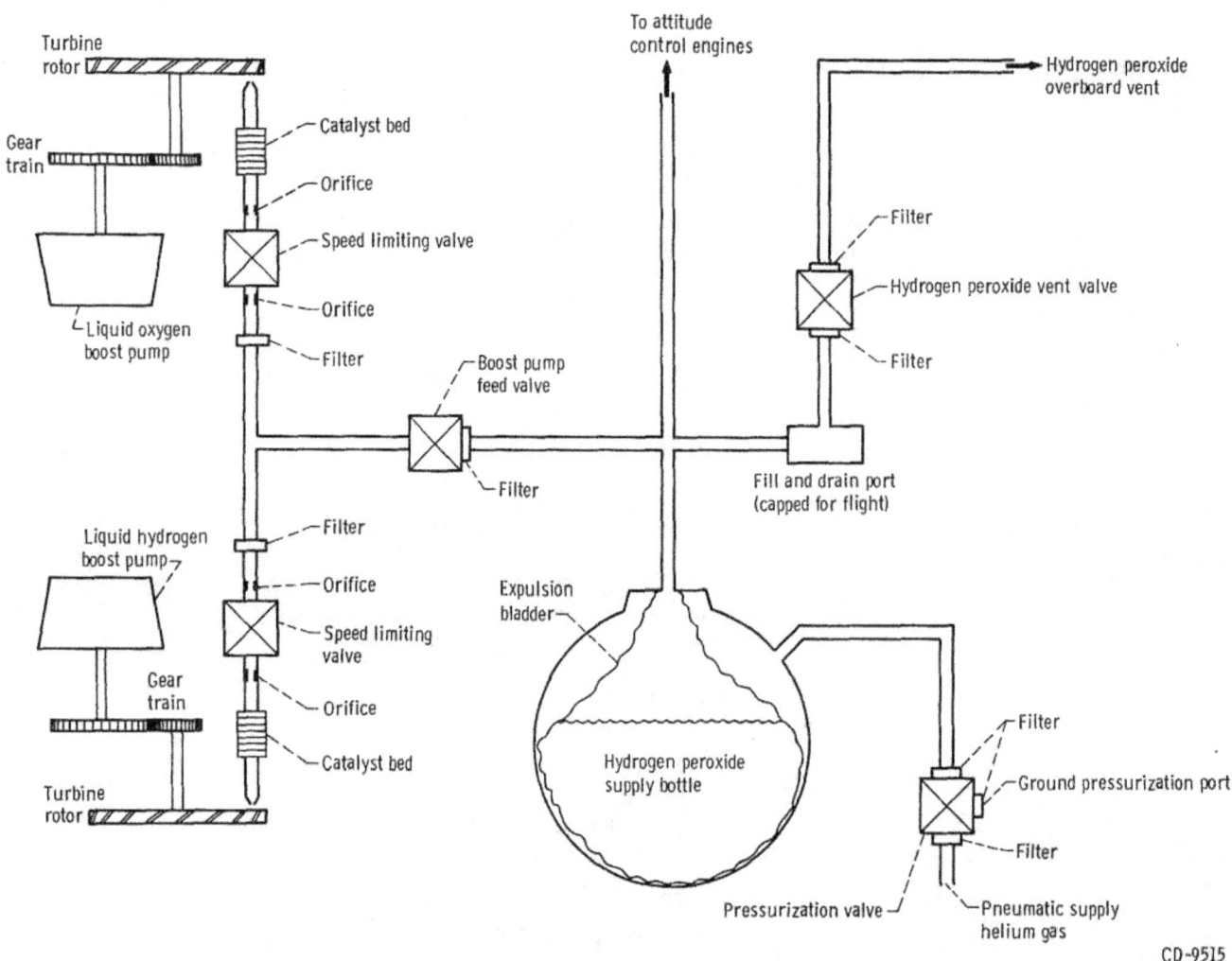

Turbine rotor

Gear train

Liquid oxygen boost pump

Catalyst bed

Orifice

Speed limiting valve

Orifice

Filter

Boost pump feed valve

Filter

Liquid hydrogen boost pump

Filter

Orifice

Speed limiting valve

Orifice

Gear train

Catalyst bed

Turbine rotor

To attitude control engines

Hydrogen peroxide overboard vent

Filter

Hydrogen peroxide vent valve

Filter

Fill and drain port (capped for flight)

Expulsion bladder

Hydrogen peroxide supply bottle

Filter

Ground pressurization port

Filter

Pressurization valve

Pneumatic supply helium gas

CD-9515

Figure VI-4. - Schematic drawing of Centaur boost pump hydrogen peroxide supply, AC-16.

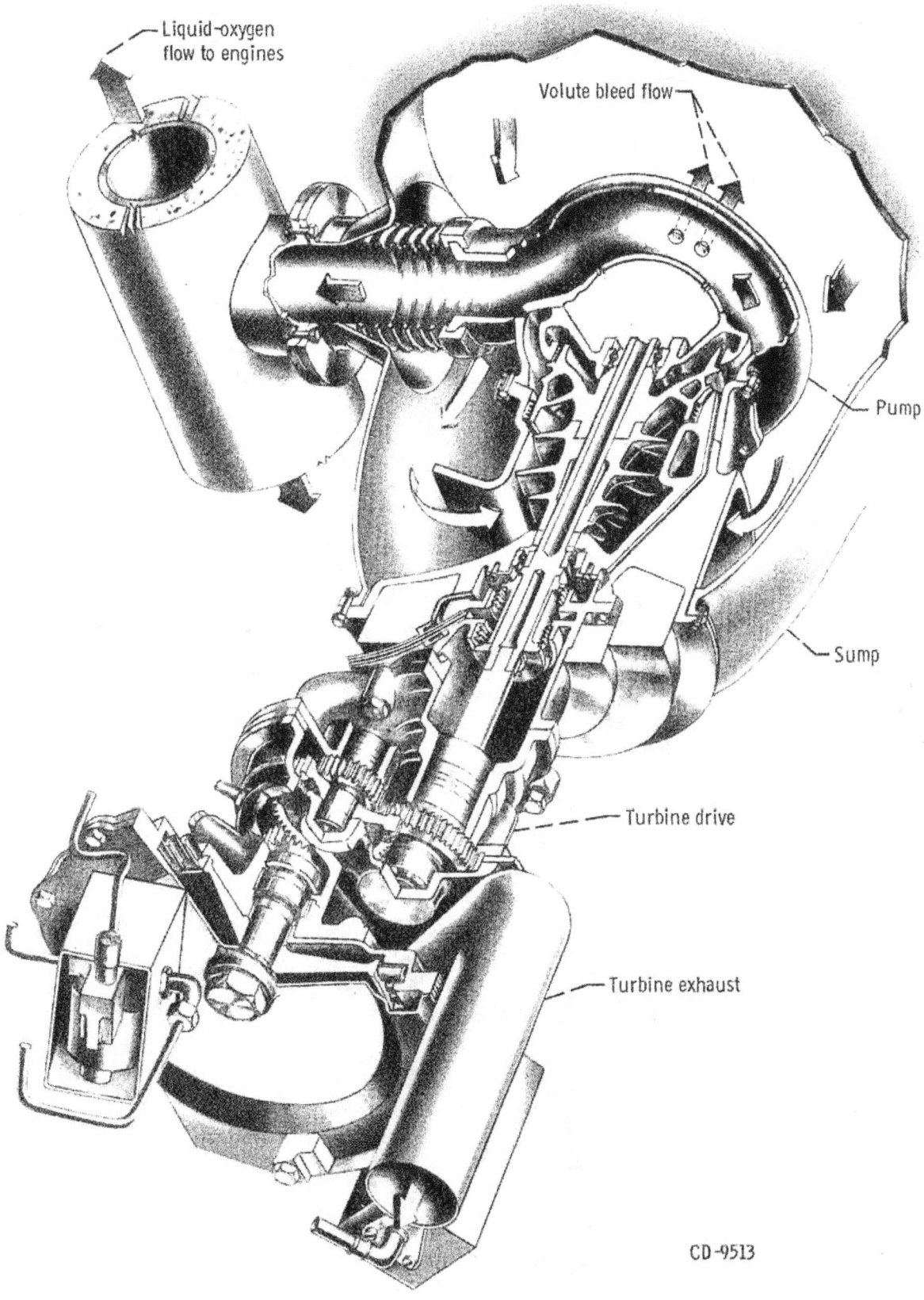

Figure VI-5. - Centaur liquid-oxygen boost pump and turbine cutaway, AC-16.

Liquid-oxygen flow to engines

Volute bleed flow

Pump

Sump

Turbine drive

Turbine exhaust

CD-9513

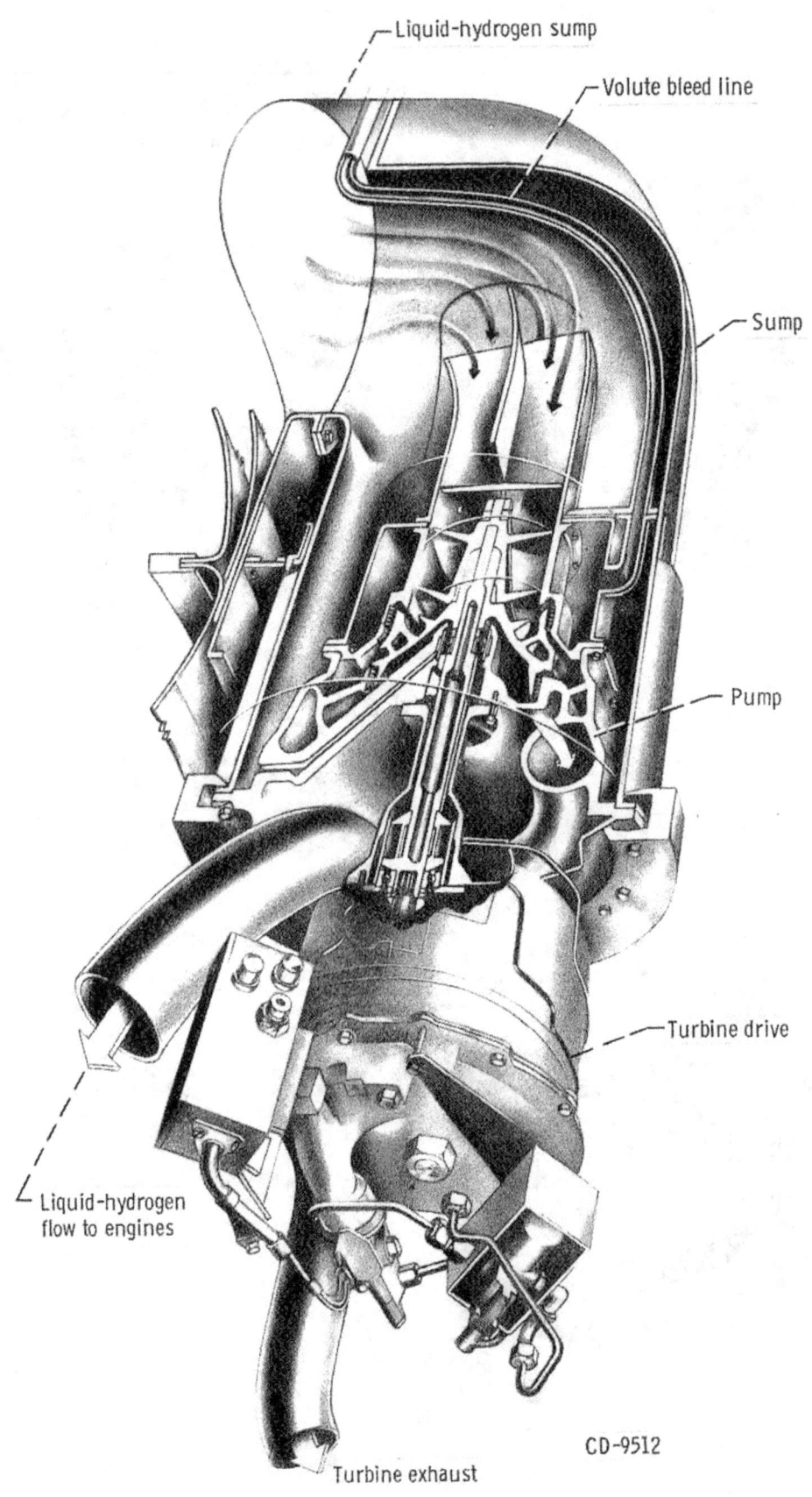

Liquid-hydrogen sump

Volute bleed line

Sump

Pump

Turbine drive

Liquid-hydrogen flow to engines

Turbine exhaust

CD-9512

Figure VI-6. - Centaur liquid-hydrogen boost pump and turbine cutaway, AC-16.

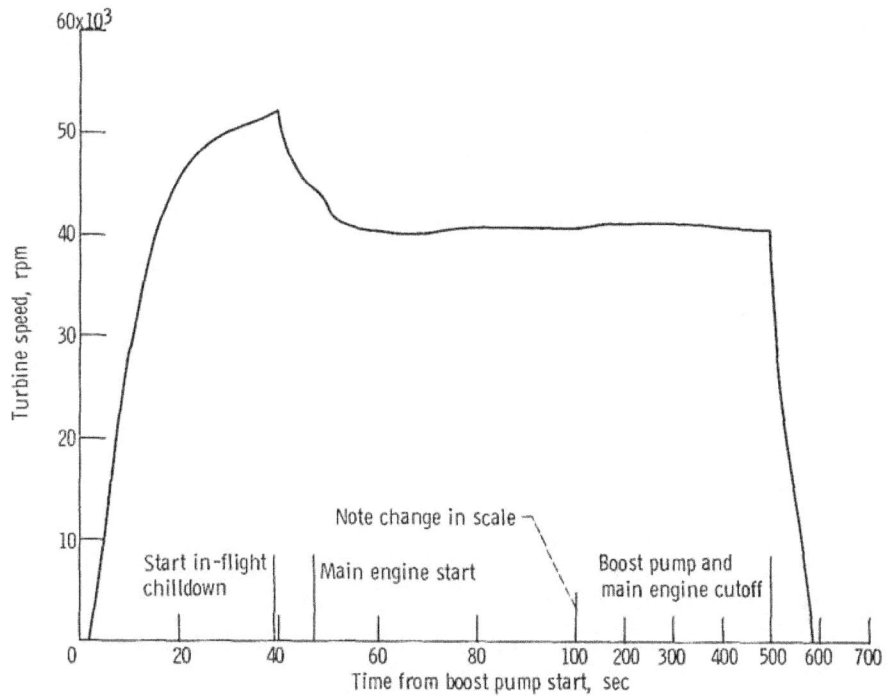

Figure VI-7. - Centaur fuel boost pump turbine speed, AC-16.

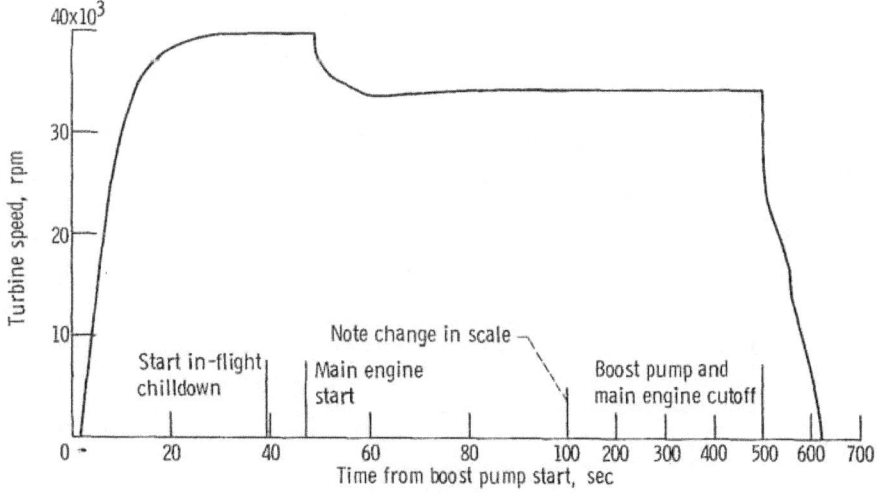

Figure VI-8. - Centaur oxidizer boost pump turbine speed, AC-16.

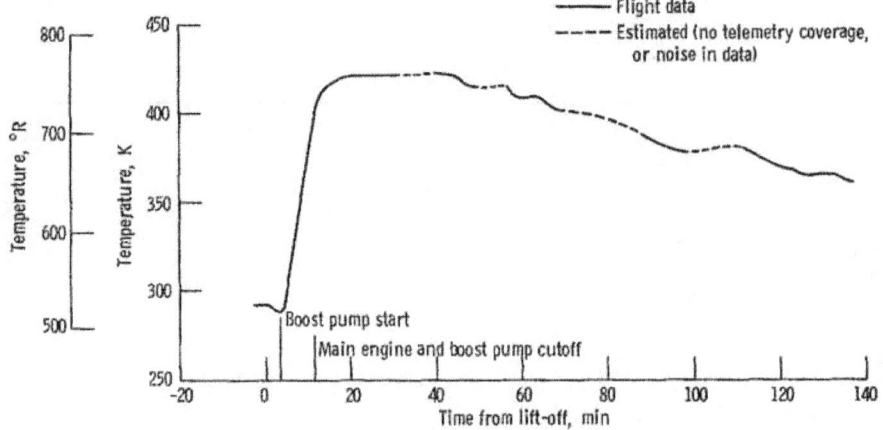

Figure VI-9. – Centaur fuel boost pump turbine bearing temperature, AC-16.

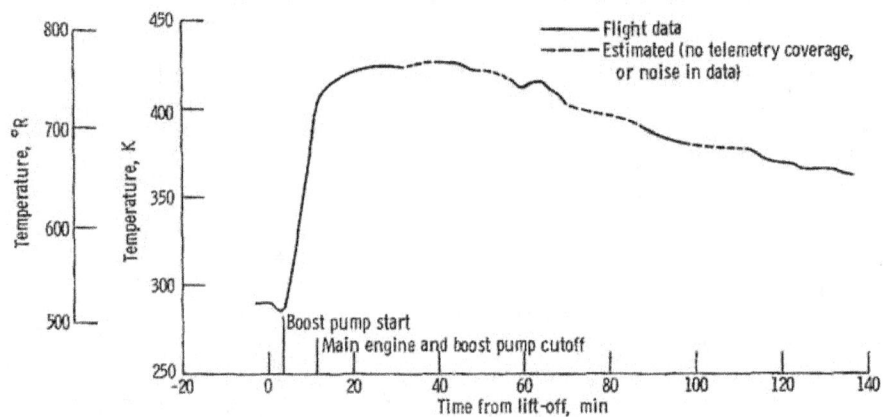

Figure VI-10. – Centaur oxidizer boost pump turbine bearing temperature, AC-16.

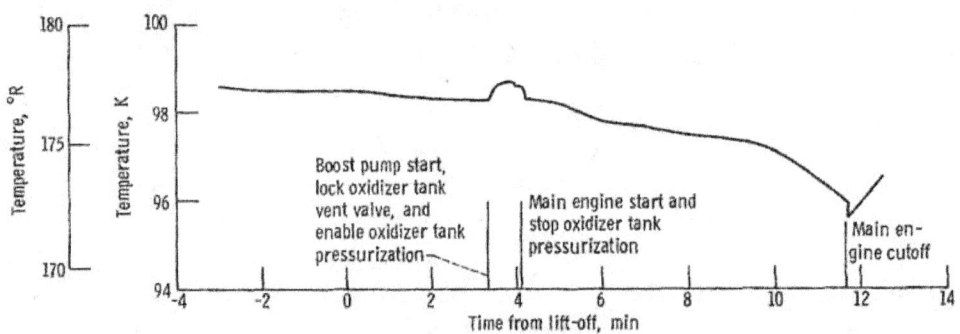

Figure VI-11. – Centaur liquid-oxygen temperature at boost pump inlet, AC-16.

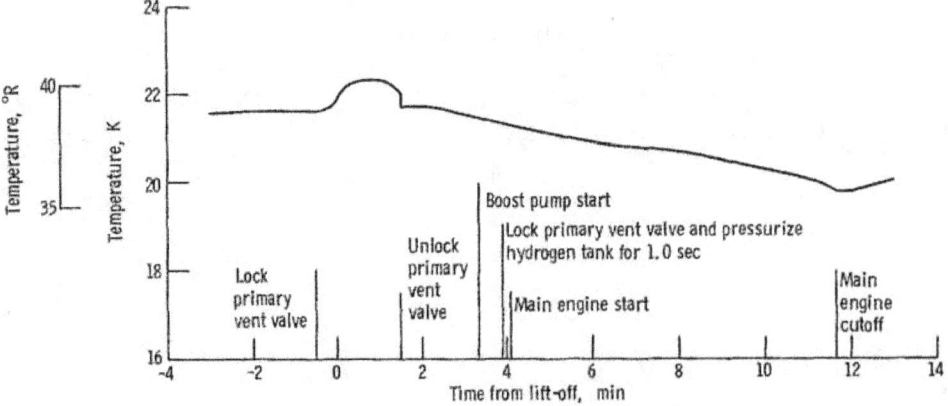

Figure VI-12. – Centaur liquid-hydrogen temperature at boost pump inlet, AC-16.

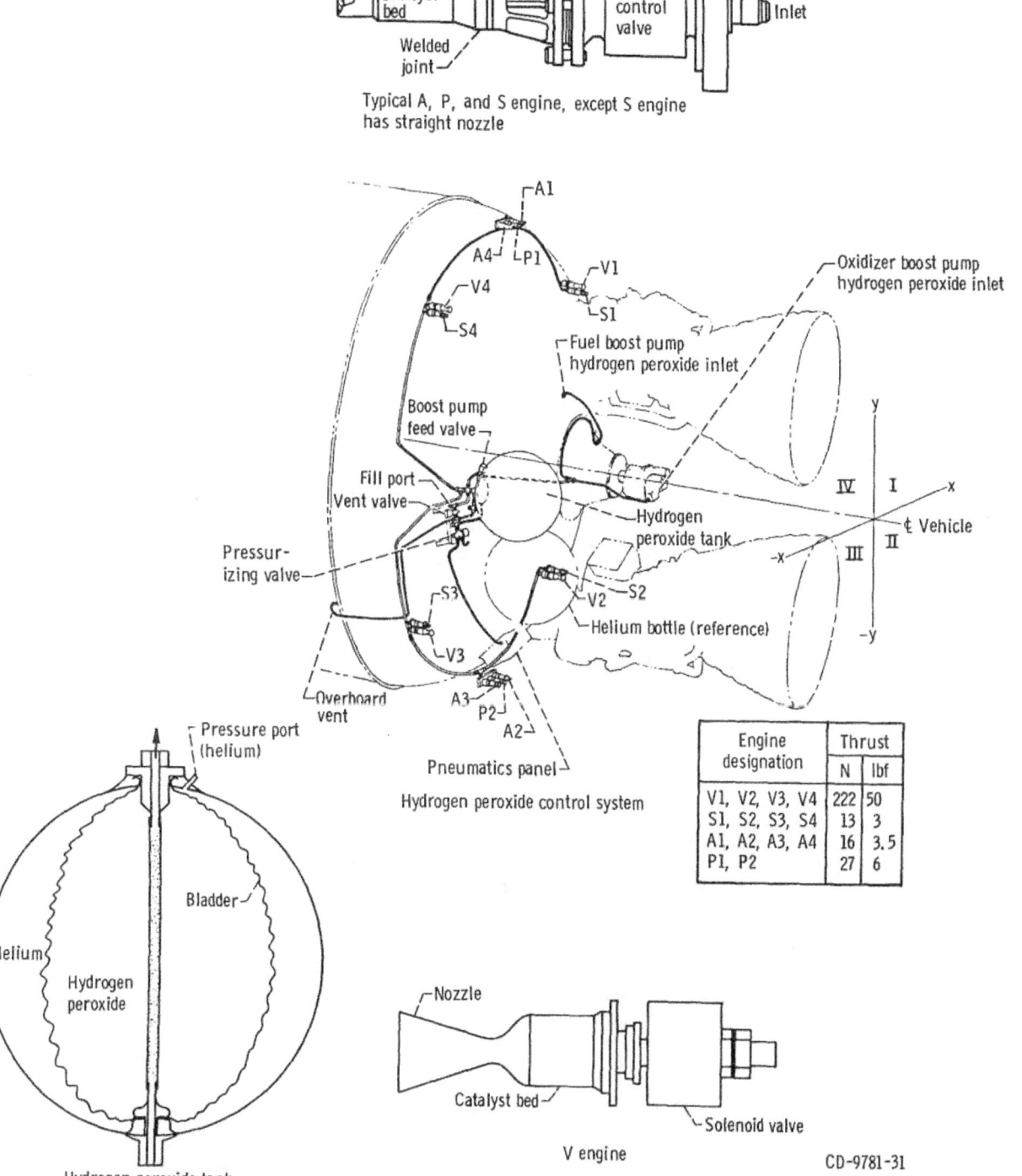

Nozzle
Heat barrier
Catalyst bed
Solenoid control valve
Inlet
Welded joint

Typical A, P, and S engine, except S engine has straight nozzle

A1
A4
P1
V1
V4
S1
S4

Oxidizer boost pump hydrogen peroxide inlet

Fuel boost pump hydrogen peroxide inlet

Boost pump feed valve

Fill port
Vent valve

Hydrogen peroxide tank

Pressurizing valve

S3
V2
S2
V3
Helium bottle (reference)

Overboard vent
A3
P2
A2

Pneumatics panel

Hydrogen peroxide control system

y
IV I
x
III II
-x
-y
¢ Vehicle

Engine designation	Thrust	
	N	lbf
V1, V2, V3, V4	222	50
S1, S2, S3, S4	13	3
A1, A2, A3, A4	16	3.5
P1, P2	27	6

Pressure port (helium)
Bladder
Helium
Hydrogen peroxide

Hydrogen peroxide tank

Nozzle
Catalyst bed
Solenoid valve

V engine

CD-9781-31

Figure VI-13. - Hydrogen peroxide system isometric, AC-16.

51

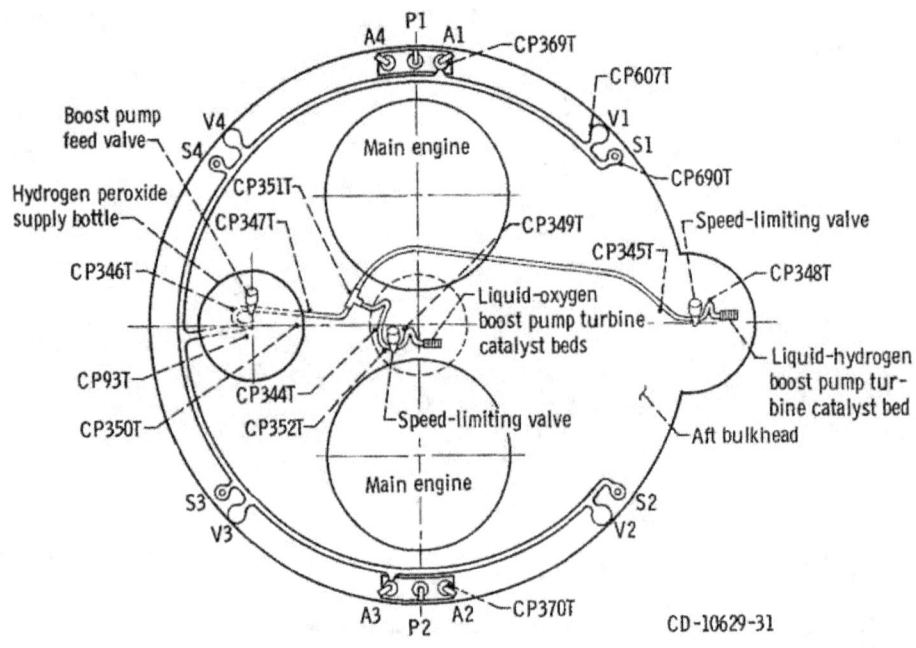

Figure VI-14. - Hydrogen peroxide system instrumentation, AC-16.
View looking forward; the A engines are positioned approximately
25° outboard from horizontal plane.

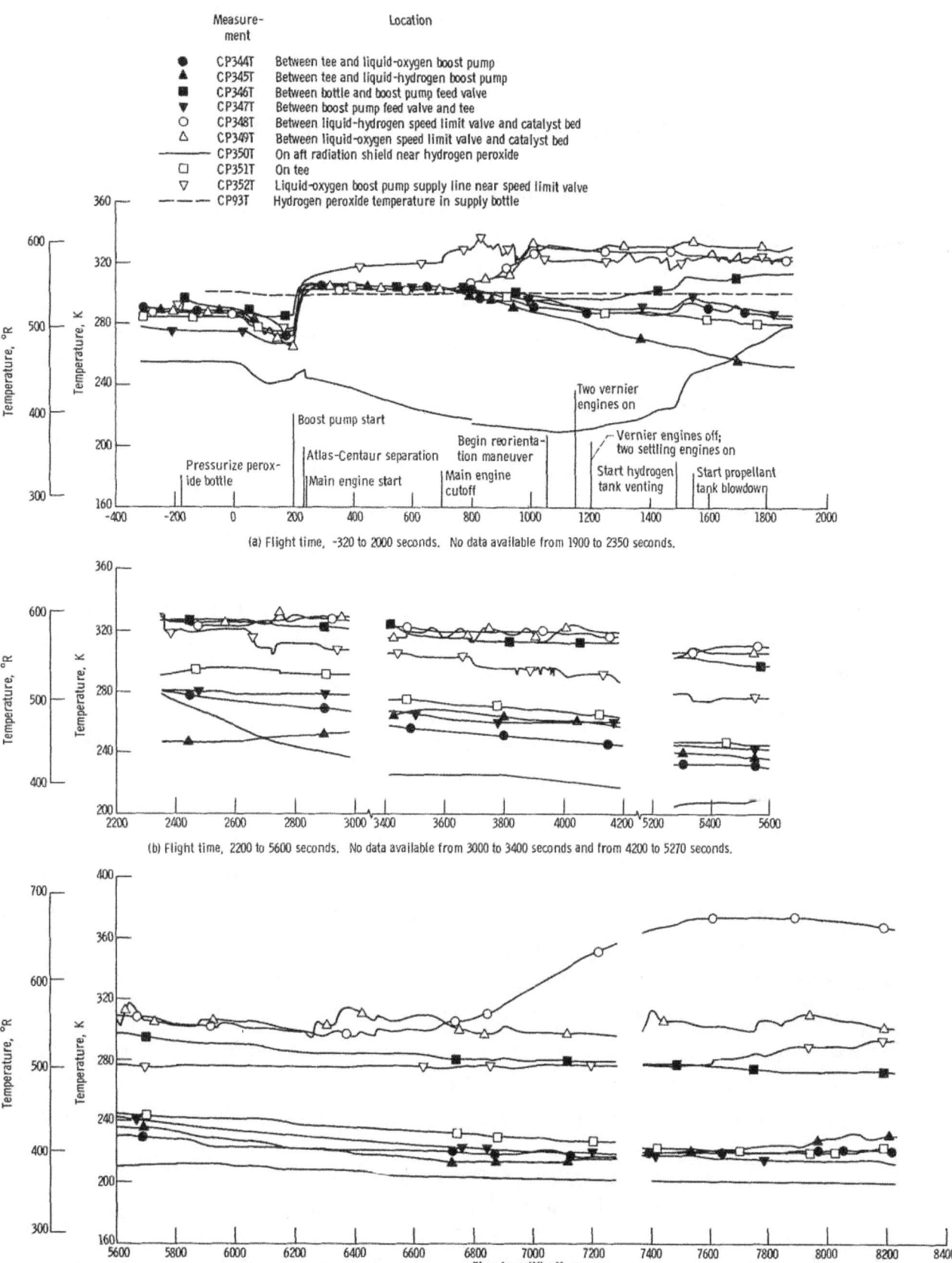

Measurement		Location
●	CP344T	Between tee and liquid-oxygen boost pump
▲	CP345T	Between tee and liquid-hydrogen boost pump
■	CP346T	Between bottle and boost pump feed valve
▼	CP347T	Between boost pump feed valve and tee
○	CP348T	Between liquid-hydrogen speed limit valve and catalyst bed
△	CP349T	Between liquid-oxygen speed limit valve and catalyst bed
——	CP350T	On aft radiation shield near hydrogen peroxide
□	CP351T	On tee
▽	CP352T	Liquid-oxygen boost pump supply line near speed limit valve
— —	CP93T	Hydrogen peroxide temperature in supply bottle

(a) Flight time, -320 to 2000 seconds. No data available from 1900 to 2350 seconds.

(b) Flight time, 2200 to 5600 seconds. No data available from 3000 to 3400 seconds and from 4200 to 5270 seconds.

(c) Flight time, 5600 to 8230 seconds. No data available from 7280 to 7370 seconds.

Figure VI-15. - Centaur boost pump hydrogen peroxide supply line temperatures, AC-16.

53

PROPELLANT LOADING AND PROPELLANT UTILIZATION

by Richard C. Kalo

Level Indicating System for Propellant Loading

System description. - The Atlas propellant level indicating system consists of a portable sight gage assembly for RP-1 fuel (kerosene) loading and platinum hot-wire-type sensors for oxidizer (liquid oxygen) loading.

The fuel loading levels are determined by visual observation of the sight gage assembly, which is connected to the fuel probe by two temporary sense lines. After tanking, the fuel sight gage assembly and sense lines are removed, and the connecting points on the vehicle are secured for flight.

The liquid-oxygen loading levels are determined from liquid sensors located at discrete points in the oxidizer (liquid oxygen) tank. The sensing elements are the hot-wire type made with platinum wire (0.0025-cm (0.001-in.) diam), which has a linear resistance-temperature coefficient. The sensors are supplied with a nearly constant current of approximately 200 milliamperes; the voltage drop across a sensor reflects the resistance value of the sensor. When uncovered, the wire has a high resistance and therefore a high voltage drop. When covered (immersed in a cryogenic fluid), it has a low resistance and low voltage drop. The control unit amplifies a change in voltage level and applies this signal to an electronic trigger circuit. When a sensor is wetted, a control relay is deenergized, and a signal is sent to the propellant loading operator.

The Centaur propellant level indicating system (fig. VI-16) utilizes platinum hot-wire level sensors in both the liquid-oxygen and liquid-hydrogen tanks. These sensors are similar in operation to those used in the Atlas liquid-oxygen tank.

System performance. - Atlas and Centaur propellant loading was satisfactorily accomplished. The weight of the Atlas fuel (RP-1) tanked was calculated to be 37 885 kilograms (83 457 lbm) based on a density of 800 kilograms per cubic meter (49 lb/ft^3). The weight of the Atlas liquid oxygen tanked was calculated to be 85 586 kilograms (188 687 lbm) based on a density of 1100 kilogram per cubic meter (69.29 lb/ft^3).

The calculated Centaur propellant weights at lift-off were 2374 kilograms (5242 lbm) ±3 percent of liquid hydrogen and 11 540 kilograms (25 476 lbm) ±1.5 percent of liquid oxygen. Data used to calculate these propellant weights are presented in table VI-VI.

Atlas Propellant Utilization System

System description. - The Atlas propellant utilization system (fig. VI-17) consists of two mercury manometer assemblies, a computer-comparator, a hydraulically actua-

54

ted propellant utilization fuel valve, sense lines, and associated electrical harnessing. The system is used to ensure nearly simultaneous depletion of the propellants and minimum propellant residuals at sustainer engine cutoff. This is accomplished by controlling the propellant mixture ratio (oxidizer flowrate to fuel flowrate) to the sustainer engine. During flight, the manometers sense propellant head pressures which are indicative of propellant mass. The mass ratio is then compared to a reference ratio (at lift-off the ratio is 2.27) in the computer-comparator. If needed, a correction signal is sent to the propellant utilization valve controlling the main fuel flow to the sustainer engine. The oxidizer flow is regulated by the head suppression valve. This valve senses propellant utilization valve movement and moves in a direction opposite to that of the propellant utilization valve. This opposite movement thus alters propellant mixture ratio to maintain constant propellant mass flow to the engine.

System performance. - The Atlas propellant utilization system operation was satisfactory. The propellant utilization system fuel valve angles during flight and the predicted valve angles are shown in figure VI-18. The valve operates at the center position for the first 13 seconds of flight because the error signal to the valve is grounded during this period to eliminate abnormal system behavior. The valve was in the fuel rich position from T + 13 seconds to approximately T + 67 seconds, and then operated in the oxygen rich position reaching the oxidizer rich stop by T + 140 seconds. The valve remained at the oxidizer rich stop until T + 169 seconds, and then operated alternately in fuel rich and oxidizer rich positions. After T + 215 seconds the valve was on the oxidizer rich stop and remained there to ensure sustainer engine cutoff by usable-liquid-oxygen depletion.

Atlas propellant residuals. - The residual propellants above the sustainer engine pump inlets at sustainer engine cutoff were calculated to be 236 kilograms (520 lbm) of liquid oxygen and 151 kilograms (336 lbm) of RP-1. These residuals were calculated by using the time the head sensing port uncovers as a reference. Calculation of the propellants consumed, from the time the port uncovers to sustainer engine cutoff, considered the effect of flow-rate decay for a liquid-oxygen depletion.

Centaur Propellant Utilization System

System description. - The Centaur propellant utilization system (fig. VI-19) is used during flight to control the ratio of propellants consumed by the main engines and to provide minimum deviation from calculated weights of usable propellant residuals. The probes (sensors) of the propellant utilization system are also used during tanking to indicate propellant levels within the range of these probes. In flight, the mass of propellant in each tank is sensed by a capacitance probe and compared in a bridge balancing circuit. If the mass ratio of propellants remaining in the tanks varies from the pre-determined value (oxidizer to fuel ratio, 5 to 1), an error signal is sent to the proportional servopositioners which control the liquid-oxygen flow control valves (one on each engine). When the mass ratio is greater than 5 to 1, the liquid-oxygen flow is increased to return the ratio to 5 to 1. When the ratio is less than 5 to 1, the liquid-oxygen flow is decreased. The sensing probes do not extend to the top of the tanks, and therefore are not used for control until after the probes are uncovered, at approximately 90 seconds after Centaur main engine start. For this 90 seconds the liquid-oxygen flow control valves are maintained at a propellant mixture ratio of approximately 5 to 1. The valves are also commanded to the null position at an approximate 5 to 1 propellant mixture ratio 15 seconds before Centaur main engine cutoff because the probes do not extend to the bottom of the tanks, and system control is lost when the liquid level depletes below the bottom of the probe.

System performance. - All prelaunch checks and calibrations of the propellant utilization system were within specifications. The in-flight operation of the system was satisfactory during Centaur main engine firing. The liquid-oxygen flow control valve angles during the engine firings are shown in figures VI-20 and VI-21.

The liquid-oxygen probe uncovered at main engine start plus 96.5 seconds. The liquid-hydrogen probe also uncovered 8.5 seconds later. The vehicle programmer commanded the valves to begin controlling at main engine start plus 89.6 seconds. The valves then moved to the oxygen rich stop and remained there for approximately 7.5 seconds. During this time, the system compensated for an excess of 40.4 kilograms (89 lbm) of liquid oxygen. The system had been programmed to compensate for 46.7 kilograms (103 lbm) of excess liquid oxygen by the propellant utilization system bias to ensure liquid-oxygen depletion. The difference between the actual correction made and the programmed value can be attributed to the following:

(1) Engine consumption rate error accumulated during the first 90 seconds of engine firing

(2) Propellant loading tolerance

Propellant residuals. - The propellant residuals were calculated by using data obtained from the propellant utilization system. The propellant residuals remaining at Centaur main engine cutoff were calculated by using the times that the propellant levels passed the bottoms of the probes as reference points. The residuals are listed in the following table:

	Units	Liquid propellant	
		Hydrogen	Oxygen
Total propellants	kg	86.3	265
	lbm	190.0	584
Usable propellants	kg	54.0	234
	lbm	119.0	515
Firing time remaining to depletion	sec	10.6	9.15

TABLE VI-VI. - CENTAUR PROPELLANT LOADING, AC-16

Quantity or event	Units	Propellant tank	
		Hydrogen	Oxygen
Amount of sensor required to be wet:			
At T - 90 seconds	percent	99.8	---------
At T - 75 seconds	percent	--------	100.2
Sensor location (vehicle station number)	-------	174.99	373.16
Tank volume at sensor[a]	m^3	35.6	12.9
	ft^3	1256.69	370.94
Ullage volume at sensor	m^3	0.32	0.23
	ft^3	11.22	6.58
Liquid-hydrogen sensor 99.8 percent uncovered at -	sec	T - 82	---------
Liquid-oxygen sensor 100.2 percent uncovered at -	sec	--------	T - 0
Ullage pressure at time sensors uncovered, absolute	N/cm^2	15.3	21.3
	psi	22.3	30.9
Density at time sensors uncovered[b]	kg/m^3	66.96	1099
	lbm/ft^3	4.185	68.68
Propellant weight in tank when sensor uncovered	kg	2382	11 540
	lbm	5259	25 476
Liquid-hydrogen boiloff prior to vent valve close[c]	kg	7.95	---------
	lbm	17.49	---------
Liquid-oxygen boiloff prior to lift-off[c]	kg	--------	0
	lbm	--------	0
Ullage volume at lift-off	m^3	0.374	0.23
	ft^3	13.2	6.58
Weight at lift-off[d]	kg	2374	11 540
	lbm	5242	25 476

[a]Volumes include 0.05-m^3 (1.85-ft^3) liquid oxygen and 0.72-m^3 (2.53-ft^3) liquid hydrogen for lines from boost pumps to engine turbopump inlet valves.

[b]Liquid-hydrogen density taken from ref. 2. Liquid-oxygen density taken from ref. 3.

[c]Boiloff rates determined from tanking test to be 0.0.15 kg/sec (0.33 lbm/sec) for liquid hydrogen and 0.14 kg/sec (0.29 lbm/sec) for liquid oxygen.

[d]Propellant loading accuracy: hydrogen, ±3.2 percent; oxygen, ±1.5 percent.

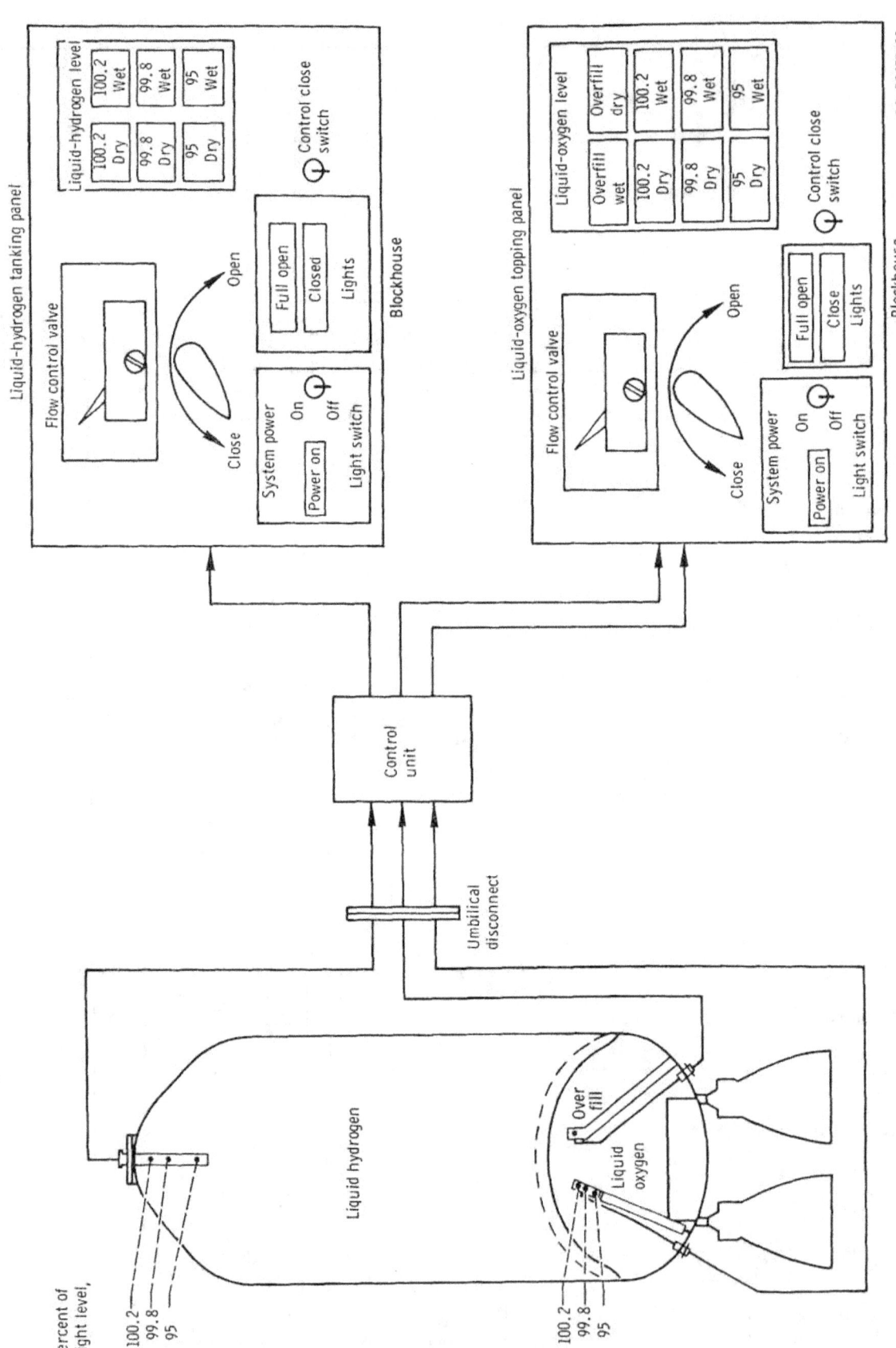

Figure VI-16. – Propellant level indicating system for Centaur propellant loading, AC-16.

CD-10457-31

58

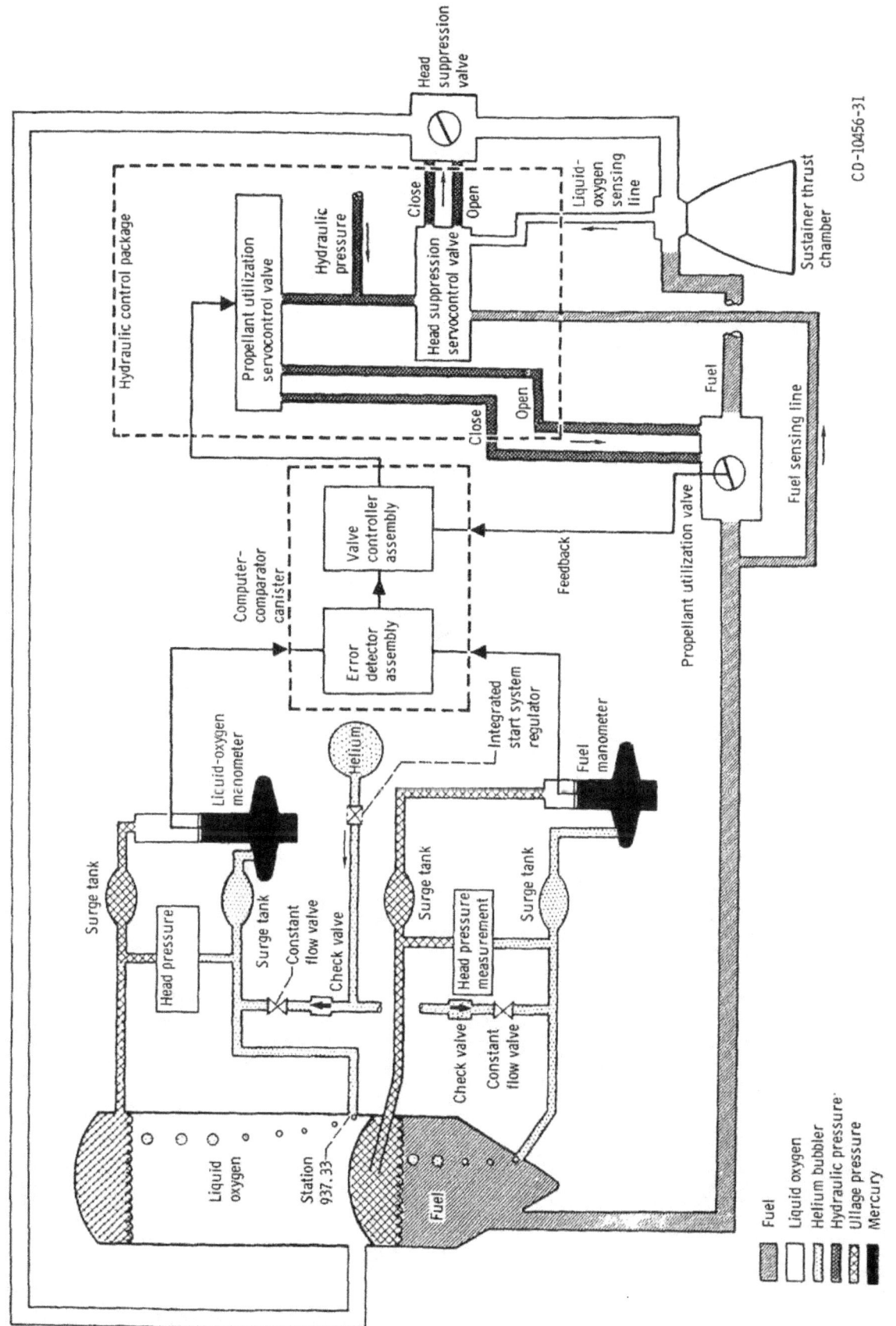

Figure VI-17. – Atlas propellant utilization system, AC-16.

CD-10456-31

59

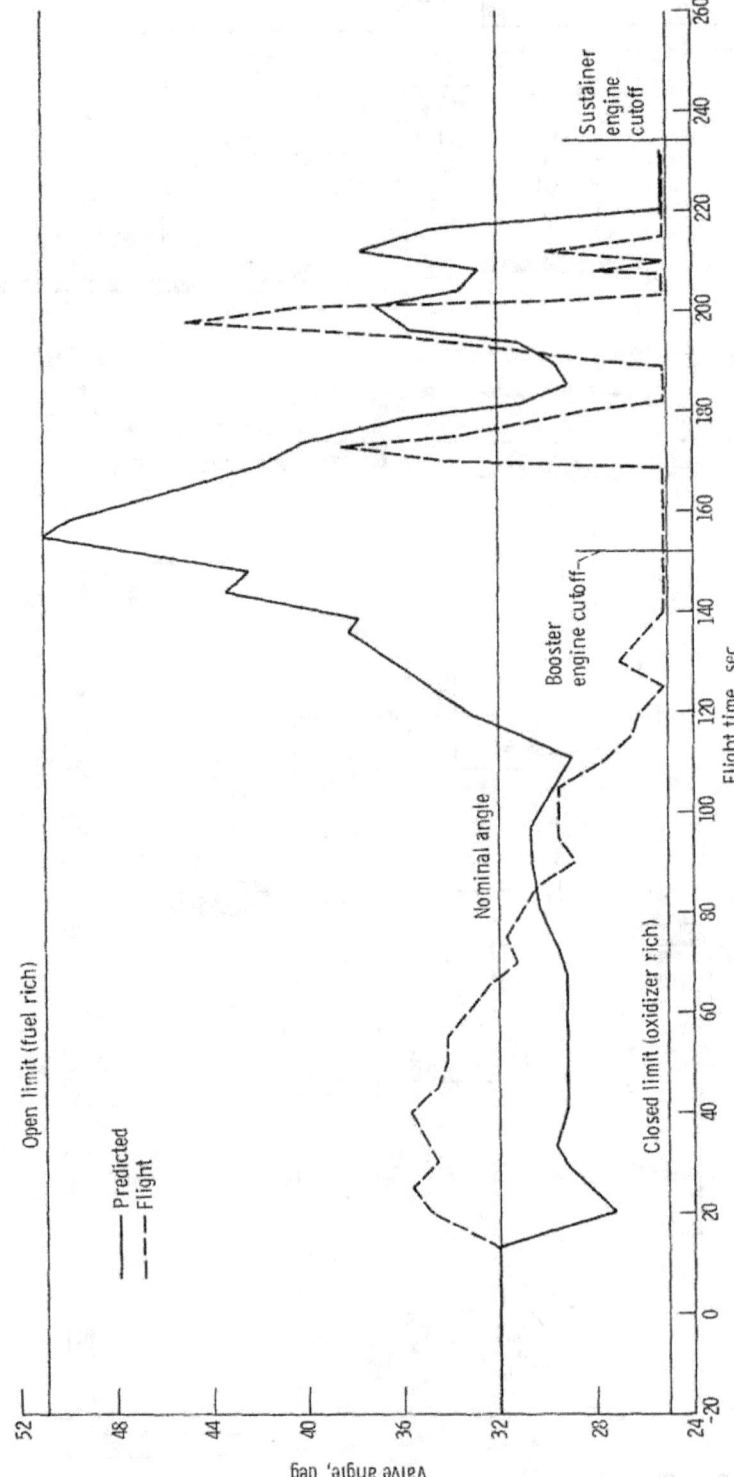

Figure VI-18. - Atlas propellant utilization fuel valve angles, predicted and flight data, AC-16. Valve angle limits: +15 percent O/F ratio, 25.1°; nominal O/F ratio, 32.1°; -15 percent O/F ratio, 50.9°.

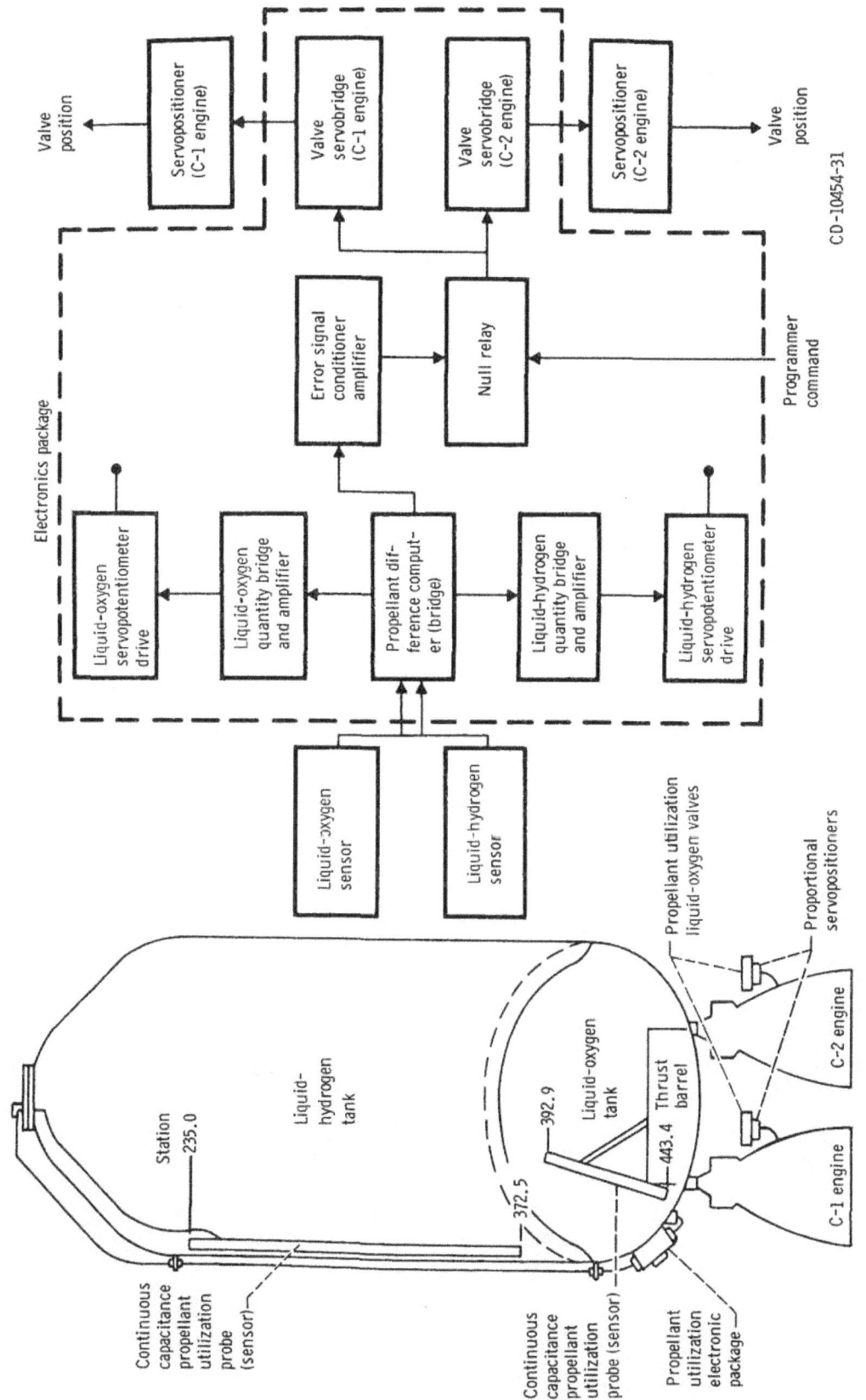

Figure VI-19. – Centaur propellant utilization system, AC-16.

CD-10454-31

61

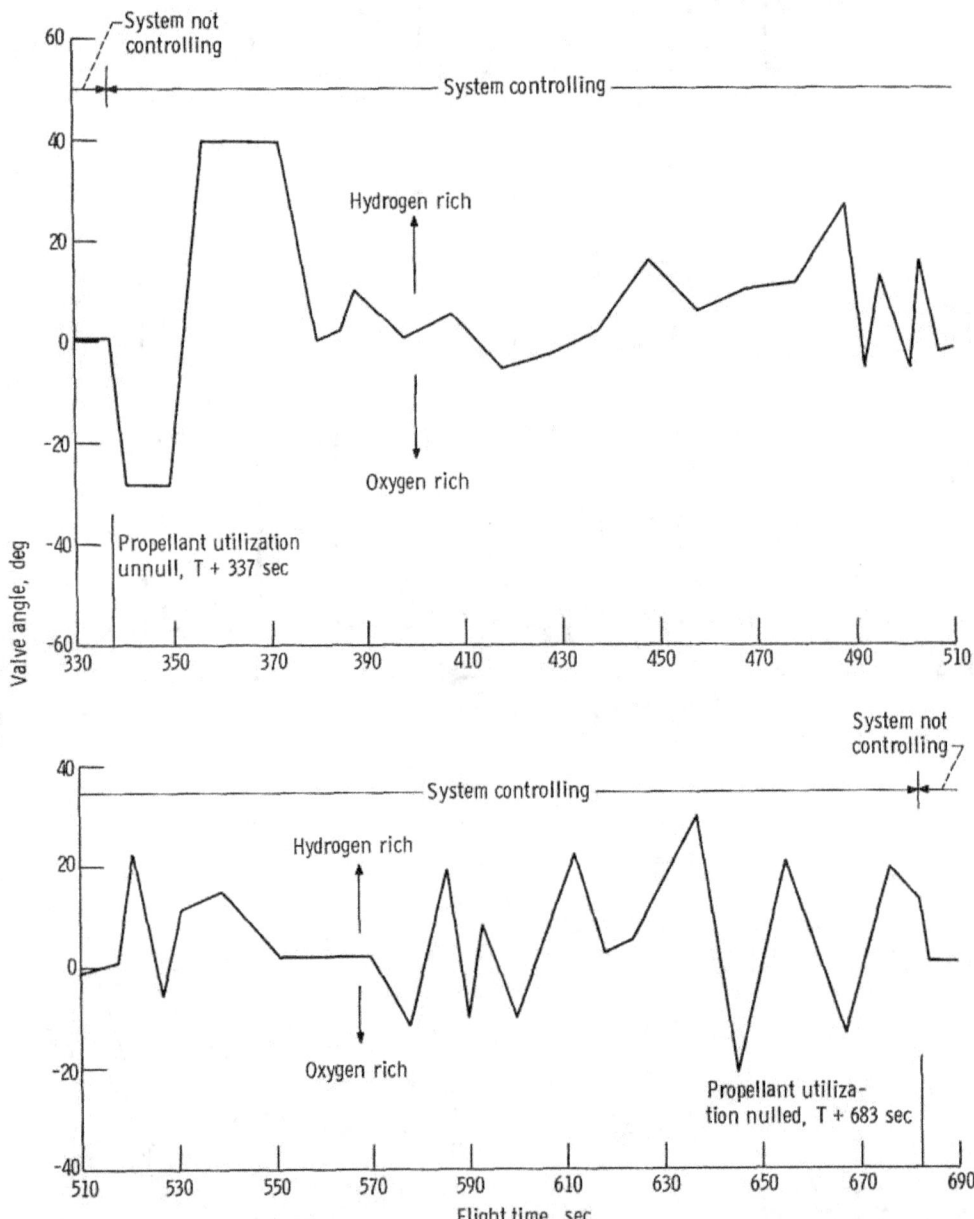

Figure VI-20. - Centaur propellant utilization valve angles for C-1 engine, AC-16. Main engine start, T + 247.5 seconds; main engine cutoff, T + 698 seconds.

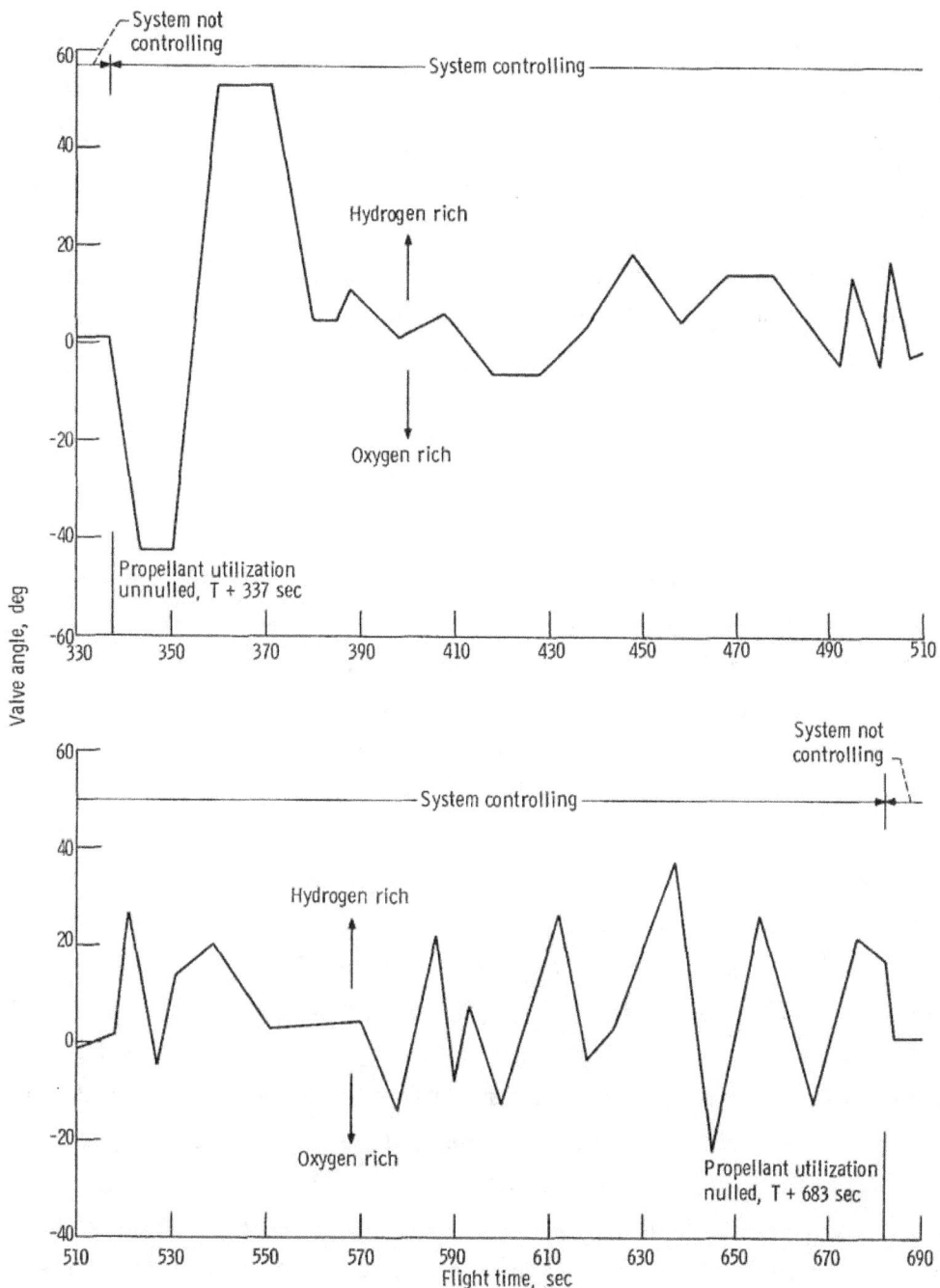

Figure VI-21. - Centaur propellant utilization valve angles for C-2 engine, AC-16. Main engine start, T + 247.5 seconds; main engine cutoff, T + 698 seconds.

PNEUMATIC SYSTEMS

by Eugene J. Fourney and Merle L. Jones

Atlas

System description. - The Atlas pneumatic system (fig. VI-22) supplies helium gas for tank pressurization and for various vehicle functions. The system is comprised of three independent subsystems: propellant tank pressurization, engine control, and booster section jettison.

Propellant tank pressurization subsystem: This subsystem is used to maintain propellant tank pressures at required levels to support the pressure-stabilized tank structure, and to satisfy the inlet pressure requirements of the engine turbopumps. In addition helium gas is supplied from the fuel tank pressurization line to pressurize the hydraulic reservoirs and turbopump lubricant storage tanks. The subsystem consists of eight shrouded helium storage bottles, a heat exchanger, and fuel and oxidizer tank pressure regulators and relief valves.

The eight shrouded helium storage bottles with a total capacity of 967 000 cubic centimeters (59 000 in.^{3}) are mounted in the jettisonable booster engine section. The bottle shrouds are filled with liquid nitrogen during prelaunch operations to chill the helium and thus provide a maximum storage capacity at about 2070 newtons per square centimeter (3000 psi). The liquid nitrogen drains from the shrouds at lift-off. During flight the cold helium passes through a heat exchanger located in the booster engine turbine exhaust duct and is heated before being supplied to the tank pressure regulators. The propellant tank pressurization subsystem pressurization control is switched from the ground to the airborne system at about T - 60 seconds. Airborne regulators are set to control fuel tank gage pressure between 44.1 and 46.2 newtons per square centimeter (64 and 67 psi) and the oxidizer tank pressure between 19.58 and 24.13 newtons per square centimeter (28.4 and 35.0 psi). From approximately T - 60 seconds to T + 20 seconds, the liquid-oxygen regulator sense line is biased by a helium "bleed" flow into the liquid-oxygen tank regulator sensing line which senses ullage pressure. The bias causes the regulator to control tank pressure at a lower level than the normal regulator setting. Depressing the liquid-oxygen tank pressure increases the differential pressure across the bulkhead between the propellant tanks to counteract the launch transient loads that act in a direction to cause bulkhead reversal. At T + 20 seconds the bias is removed by closing explosively actuated valves, and the ullage pressure in the liquid-oxygen tank increases to the normal regulator control range. The increased pressure then provides sufficient vehicle structural stiffness to withstand bending loads during the remainder of the ascent.

Pneumatic regulation of tank pressure is terminated at booster engine staging. Thereafter, the fuel tank pressure decays slowly, but the oxidizer tank pressure decay is less than the fuel tank pressure decay since it is partially sustained by liquid-oxygen boiloff.

Engine controls subsystem: The engine controls subsystem supplies helium pressure for actuation of engine control valves, for pressurization of the engine start tanks, for purging booster engine turbopump seals, and for the reference pressure to the regulators which control oxidizer flow to the gas generator. Pressure control in the system is maintained through Atlas-Centaur separation. These pneumatic requirements are supplied from a single 76 000-cubic centimeter (4650-in.^{3}) storage bottle pressurized to a gage pressure of about 2070 newtons per square centimeter (3000 psi) at lift-off.

Booster engine jettison subsystem: The booster engine jettison subsystem supplies pressure for release of the pneumatic staging latches to separate the booster engine package. A command from the Atlas flight control system opens two explosively actuated valves to supply helium pressure to the 10 piston-operated staging latches. Helium for the system is supplied by a single 14 260-cubic-centimeter (870-in.^{3}) bottle charged to a gage pressure of 2070 newtons per square centimeter (3000 psi).

System performance. - Atlas pneumatic system performance data are presented in table VI-VII. Individual subsystem performance during flight was as follows:

Propellant tank pressurization subsystem: Control of the propellant tank pressures was switched from the ground pressurization control unit to the airborne regulators at approximately T - 63 seconds. Ullage pressures were properly controlled throughout the flight.

The fuel tank pressure regulator controlled at a gage pressure of about 45.75 newtons per square centimeter (66.5 psi) until termination of pneumatic control at booster engine staging. During the sustainer phase, the fuel tank ullage pressure decreased normally and was 38.87 newtons per square centimeter (56.14 psi) at sustainer engine cutoff.

The oxidizer tank ullage pressure was steady at 21.24 newtons per square centimeter (30.8 psi) after switching from the pressurization control unit to "pneumatics internal" at approximately T - 63 seconds. The pressure decreased to 20.75 newtons per square centimeter (30.1 psi) at engine start and decreased slightly until T + 20 seconds. At T + 20 seconds the oxidizer ullage tank pressure was 20.62 newtons per square centimeter (29.9 psi). At this time the regulator sense line bias was terminated, and the pneumatic regulator increased the liquid-oxygen tank ullage pressure to 23.20 newtons per square centimeter (33.8 psi). Three seconds were required for the ullage pressure to stabilize. The liquid-oxygen tank pressure remained within the required limits until termination of pneumatic regulation at booster engine staging. After booster engine staging, the ullage pressure decreased from 23.05 to 21.51 newtons per square centimeter (33.45 to 31.19 psi) at sustainer engine cutoff.

Engine control subsystem: The booster and sustainer engine control regulators provided the required helium pressure for engine control throughout the flight.

Booster section jettison subsystem: Booster section jettison subsystem performance was satisfactory. The explosively actuated valve was opened on command, allowing high-pressure helium to actuate the 10 booster staging latches.

Centaur

System description. - The Centaur pneumatic system, which is shown schematically in figure VI-23, consists of four subsystems: propellant tank venting, propellant tank pressurization, propulsion pneumatics, and helium purge pneumatics.

Propellant tank venting subsystem: The structural stability of the propellant tanks is maintained throughout the flight by the propellant boiloff gas pressures. These pressures are controlled by a vent system on each propellant tank. Two pilot-controlled, pressure-actuated vent valves and ducting comprise the hydrogen tank vent system. The primary vent valve is fitted with a continuous-duty solenoid valve which, when energized, prevents the vent valve from relieving. The secondary hydrogen vent valve does not have the control solenoid and is always in the "unlocked" mode. The relief range of the secondary valve is above that of the primary valve, preventing overpressurization of the hydrogen tank when the primary vent valve is locked. The vented hydrogen gas is ducted overboard through a single vent. The oxygen tank vent system uses a single vent valve which is fitted with the control solenoid valve. The vented oxygen gas is ducted overboard through the interstage adapter. The duct, which remains with the Centaur after separation from the interstage adapter, is oriented to approximately aline the venting thrust vector with the vehicle center of gravity.

The vent valves are commanded to the locked mode at specific times (1) to satisfy the structural requirements of the pressure-stabilized tank, (2) to permit controlled pressure increases in the tanks to satisfy the boost pump pressure requirements, (3) to restrict venting during nonpowered flight to avoid vehicle disturbing torques, and (4) to restrict hydrogen venting to nonhazardous times. (A fire could conceivably occur during the early part of the atmospheric ascent if a plume of vented hydrogen washed back over the vehicle and if it were exposed to an ignition source. A similar hazard could occur at Atlas booster engine staging when residual oxygen envelops a large portion of the vehicle.)

Propellant tank pressurization subsystem: The propellant tank pressurization subsystem supplies helium gas in controlled quantities for in-flight pressurization, in addition to that provided by the propellant boiloff gases. It consists of a helium storage bottle, two normally closed solenoid valves and orifices, and a pressure switch assembly which senses oxygen tank pressure. The solenoid valves and orifices provide metered flow of helium to both propellant tanks for step pressurization during the main engine start sequence and to the oxygen tank at main engine cutoff. The pressure-sensing switch controls the pressurization of the oxygen tank during the main engine start sequence.

Propulsion pneumatics subsystem: The propulsion pneumatics subsystem supplies helium gas from the helium storage bottle at regulated pressures for actuation of main engine control valves and pressurization of the hydrogen peroxide storage bottle. It consists of two pressure regulators, which are referenced to ambient pressure, and two relief valves. Pneumatic pressure supplied through the engine controls regulator is used for actuation of the engine inlet valves, the engine cooldown valves, and the main fuel shutoff valve. The second regulator, located downstream of the engine controls regulator, further reduces the pressure to provide expulsion pressurization for the hydrogen peroxide storage bottle. A relief valve downstream of each regulator prevents overpressurization.

Helium purge pneumatics subsystem: A ground-airborne helium purge subsystem is used to prevent cryopumping and icing under the insulation panels and in propulsion system components. A common airborne distribution system is used for prelaunch purging from a ground helium source and postlaunch purging from an airborne helium storage bottle. This subsystem distributes helium gas for purging the cavity between the hydrogen tank and the insulation panels, the seal between the barrel section and the forward bulkhead, the propellant feedline insulation, the boost pump seal vents, the engine gearbox seal vents, the engine chilldown vent-ducts, the engine thrust chambers, and the hydraulic power packages. The umbilical charging connection for the airborne bottle can also be used to supply the purge from the ground source should an abort occur after ejection of the ground purge supply line.

System performance. - The pneumatic system performance during the flight was as follows:

Propellant tank pressurization and venting subsystems: The ullage pressures for the hydrogen and oxygen tanks during the flight are shown in figure VI-24. The hydrogen tank absolute pressure was 14.5 newtons per square centimeter (21.0 psi) at T - 28.8 seconds when the primary hydrogen vent valve was locked. The valve was locked earlier than normal on this flight to provide a greater structural capability at lift-off through an increased hydrogen tank pressure. (On most Atlas-Centaur flights the vent valve is locked at T - 8 sec.) The need for greater strength at lift-off is discussed in the section VEHICLE STRUCTURES. After vent valve lockup, the tank ullage absolute pressure increased, at an average rate of 3.51 newtons per square centimeter per minute (5.09 psi/min), to 18.0 newtons per square centimeter (26.1 psi) at T + 31 seconds. At this time the secondary vent valve relieved and regulated tank pressure until T + 90 seconds, when the primary vent valve was enabled. The tank pressure was then reduced and was regulated by the primary vent valve.

At T + 152.1 seconds the primary hydrogen vent valve was locked for 7.1 seconds during Atlas booster engine staging. Following booster engine staging the primary vent valve was enabled and allowed to regulate tank pressure. At T + 234.1 seconds the primary hydrogen vent valve was again locked, and the tank was pressurized with helium for 1 second. The tank absolute pressure increased from 13.7 to 14.7 newtons per square centimeter (19.9 to 21.3 psi). As the warm helium in the tank cooled, the absolute pressure decreased to 13.9 newtons per square centimeter (20.2 psi) at T + 246.0 seconds (Centaur main engine start). The absolute pressure at engine prestart (T + 238.0 sec) was 14.5 newtons per square centimeter (21.0 psi) (figs. VI-24(a) and (b)).

The ullage absolute pressure in the oxygen tank was 20.6 newtons per square centimeter (29.9 psi) at lift-off. After lift-off the pressure began to decrease with the increasing vehicle acceleration which suppressed the propellant boiling. At T + 90 seconds the vent valve reseated and venting ceased. The pressure then began to increase and decrease alternately with vent valve operation until Atlas booster engine cutoff. At this time the sudden reduction in the acceleration caused an increase in the liquid-oxygen boiloff and an ullage pressure rise. As thermal equilibrium was reestablished in the tank, the ullage pressure decreased.

At T + 199.2 seconds the oxygen tank vent valve was locked, and the helium pressurization of the tank began. The tank ullage absolute pressure increased to 27.4 newtons per square centimeter (39.7 psi), which was the upper limit of the pressure switch. As the warm helium gas cooled in the tank, the absolute pressure decreased to 26.2 newtons per square centimeter (38.0 psi), when the pressure switch closed, and additional helium was injected into the tank. After the second cycle, the heat input from the boost pump recirculation flow increased the boiloff and caused the pressure to increase before it reached the lower limit of the pressure switch. At engine prestart the absolute pressure was 28.4 newtons per square centimeter (41.2 psi). After engine prestart the absolute pressure decreased to 27.5 newtons per square centimeter (39.9 psi) at main engine start and decreased thereafter to the saturation value of the oxygen gas (figs. VI-24(a) and (b)).

The ullage pressures in both propellant tanks decreased normally during main engine firing. At engine cutoff the ullage absolute pressures in the hydrogen and oxygen tanks were 9.1 and 16.5 newtons per square centimeter (13.2 and 23.9 psi), respectively. At engine cutoff the oxygen tank was pressurized with helium for 90 seconds in order to preclude the possibility of the hydrogen tank pressure exceeding the oxygen tank pressure and reversing the intermediate bulkhead. During this period the oxygen tank absolute pressure increased to 19.6 newtons per square centimeter (28.4 psi). The pressure continued to increase to 20 newtons per square centimeter (29.0 psia) at T + 900 seconds and remained constant through start of retrothrust. The hydrogen tank absolute pressure increased to 18.0 newtons per square centimeter (26.1 psi) during the period between main engine cutoff and T + 1491 seconds. The secondary hydrogen vent valve then relieved and regulated tank pressure (figs. VI-24(c) and (d)).

After the start of retrothrust the hydrogen tank ullage pressure remained constant for approximately 70 seconds, indicating liquid outflow. The pressure then began to decrease, indicating either gaseous or two-phase outflow. The oxygen tank ullage pressure remained constant to loss of signal (T + 1890 sec), indicating liquid outflow (fig. VI-24(d)). At acquisition of signal (T + 2350 sec) the pressure was decreasing, indicating gaseous or two-phase outflow.

Propulsion pneumatics subsystem: The engine controls regulator and the hydrogen peroxide bottle pressure regulator maintained proper system pressure levels throughout the flight. The engine controls regulator output absolute pressure was 326 newtons per square centimeter (473 psi) at T - 0 seconds, while that of the hydrogen peroxide bottle pressure regulator was 224 newtons per square centimeter (325 psi). After lift-off both regulator output pressures decreased corresponding to the decrease in ambient pressure and remained relatively constant after the ambient pressure had decreased to zero.

Helium purge subsystem: The total helium purge flow rate to the vehicle at T - 14 seconds was 86 kilograms per hour (190 lbm/hr). The differential pressure across the insulation panels after hydrogen tanking was 0.15 newton per square centimeter (0.22 psi). The minimum allowable differential pressure required to prevent cryopumping and icing is 0.02 newton per square centimeter (0.03 psi). At T - 13.6 seconds the airborne purge system was activated, and at T - 4 seconds the ground purge was terminated. The supply of helium in the purge bottle lasted through most of the atmospheric ascent.

TABLE VI-VII. - ATLAS PNEUMATIC SYSTEM PERFORMANCE, AC-16

Parameter	Measurement number	Units	Design range	Flight values at -					
				T - 10 sec	T - 0 sec	T + 20 sec	T + 23 sec	Booster engine cutoff	Sustainer or vernier engine cutoff
Oxidizer tank ullage pressure, gage	AF1P	N/cm^2 psi	(a)	21.24 30.8	20.75 30.1	[b]20.62 29.9	[c]23.20 33.8	23.05 33.45	[d]21.51 31.19
Fuel tank ullage pressure, gage	AF3P	N/cm^2 psi	44.13 to 46.19 64.0 to 67.0	45.85 66.5	45.16 65.5	45.75 66.5	46.06 66.8	45.57 66.1	[d]38.87 56.14
Intermediate bulkhead differential pressure[e]	AF116P	N/cm^2 psi	0.345 (min.) 0.5 (min.)	14.03 20.5	14.62 21.2	12.6 18.25	9.48 13.75	17.24 25.0	17.24 25.0
Sustainer controls helium bottle pressure, absolute	AF291	N/cm^2 psi	2344 (max.) 3400 (max.)	2275 3300	2179 3160	2131 3090	2131 3090	1934 2805	1862 2700
Booster helium bottle pressure, absolute	AF246P	N/cm^2 psi	2344 (max.) 3400 (max.)	2310 3350	2186 3170	1700 2465	1634 2370	438 635	(d)
Booster helium bottle temperature	AF247T	^oR ^oF	154.67 to 139.67 (-305 to -320) prior to engine start	140.57 -319.2	139.07 -320.6	125.07 -334.6	123.67 -336.0	82.67 -377	(d)

[a]Prior to T-0 sec, 19.58 to 22.17 N/cm^2 (28.4 to 32.3 psi); after T + 23 sec, 22.06 to 24.13 N/cm^2 (32.0 to 35.0 psi).

[b]Signal from programmer to fire programmed pressure conax valves.

[c]Oxidizer tank pressure at termination of programmed pressure.

[d]Helium supply bottles jettisoned with booster at booster engine cutoff plus 3 sec.

[e]Lowest valve was 13 psi at T + 1 sec.

70

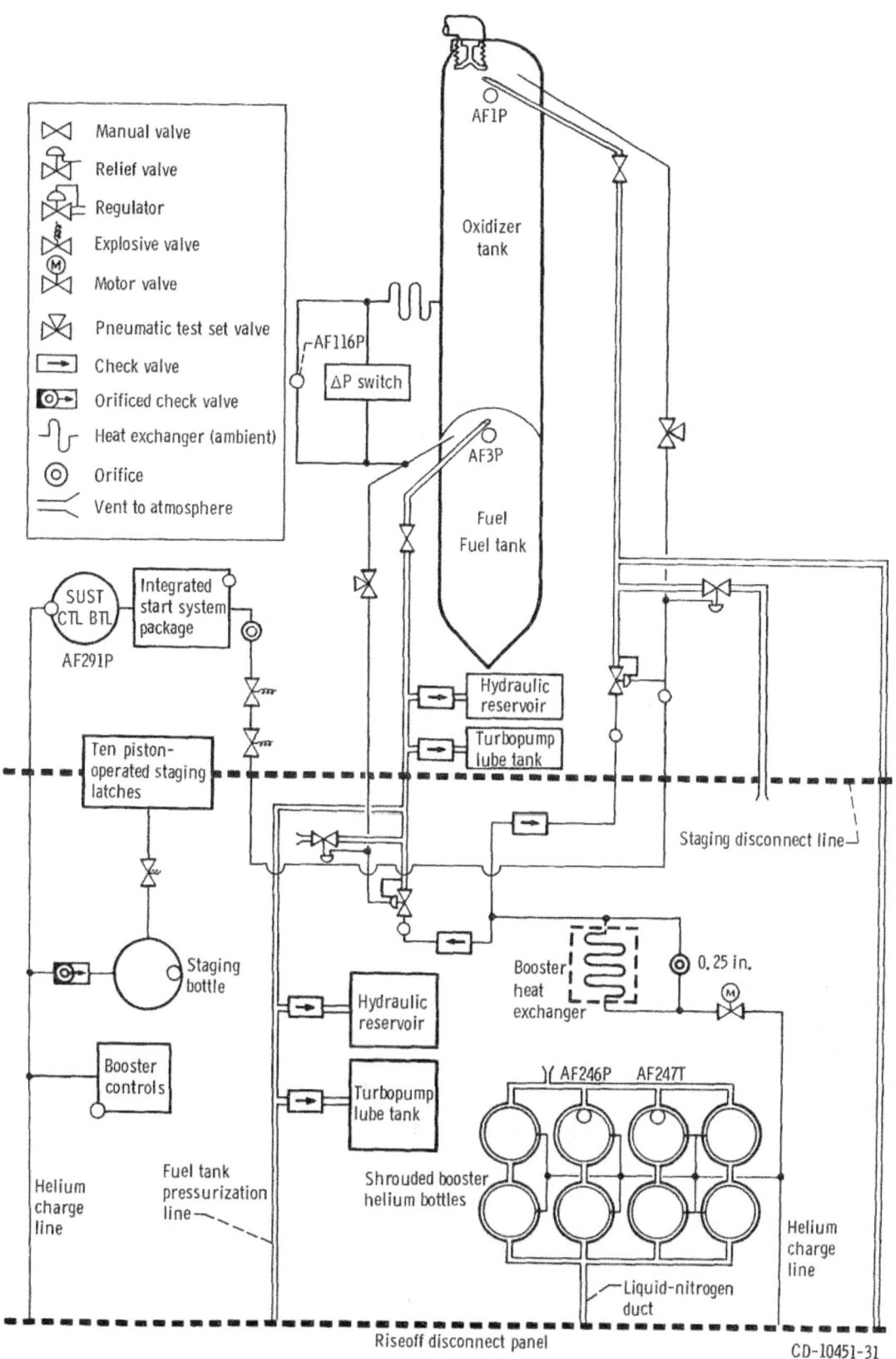

Legend:
- ⋈ Manual valve
- Relief valve
- Regulator
- Explosive valve
- Ⓜ Motor valve
- Pneumatic test set valve
- → Check valve
- ⊙→ Orificed check valve
- Heat exchanger (ambient)
- ⊙ Orifice
- Vent to atmosphere

AF1P

Oxidizer tank

AF116P

ΔP switch

AF3P

Fuel
Fuel tank

SUST CTL BTL

Integrated start system package

AF291P

Ten piston-operated staging latches

Hydraulic reservoir

Turbopump lube tank

Staging disconnect line

Staging bottle

Booster controls

Hydraulic reservoir

Turbopump lube tank

Booster heat exchanger

0.25 in.

AF246P   AF247T

Shrouded booster helium bottles

Helium charge line

Fuel tank pressurization line

Liquid-nitrogen duct

Helium charge line

Riseoff disconnect panel

CD-10451-31

Figure VI-22. - Atlas vehicle pneumatic system, AC-16.

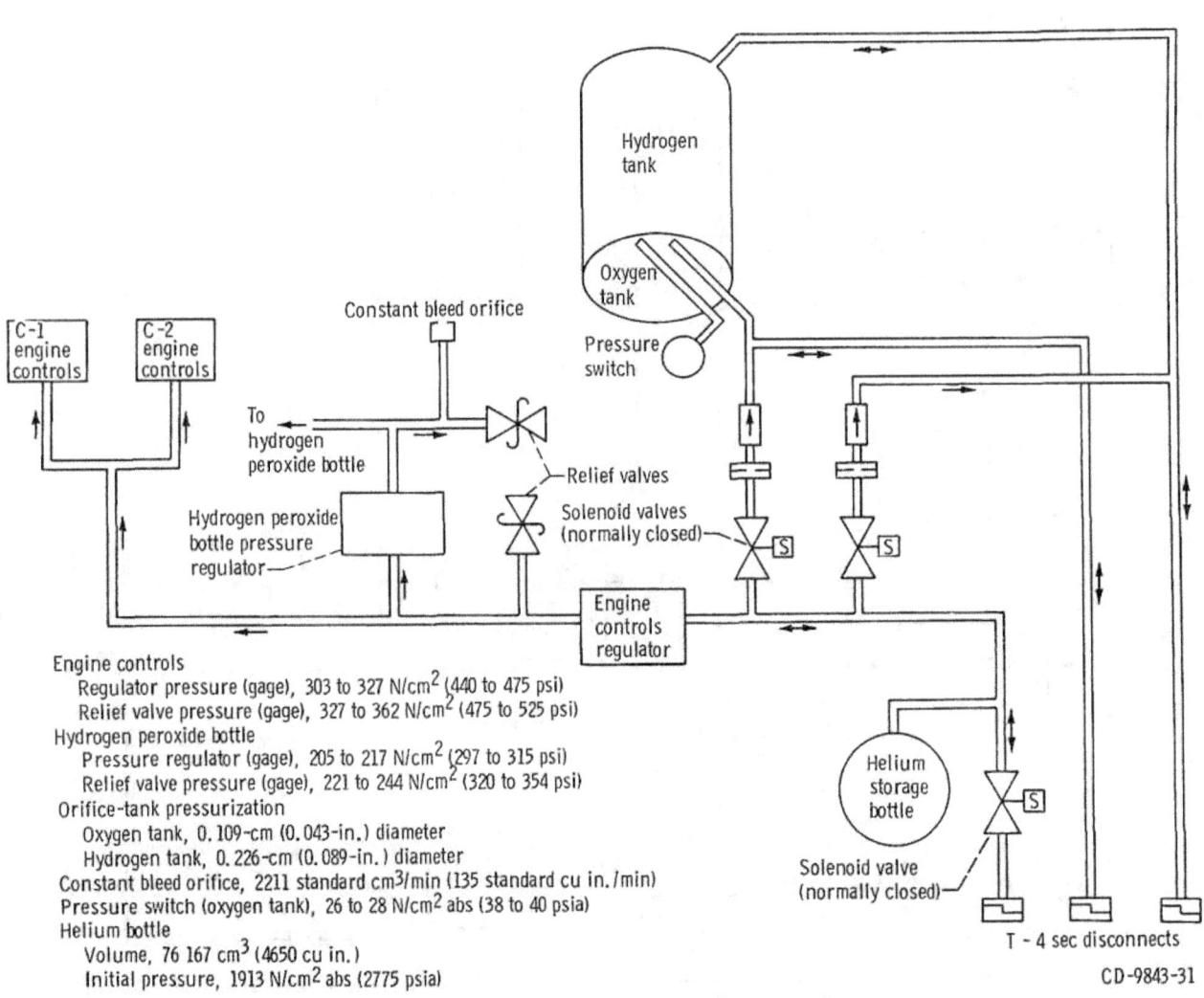

Constant bleed orifice

C-1 engine controls

C-2 engine controls

Hydrogen tank

Oxygen tank

Pressure switch

To hydrogen peroxide bottle

Relief valves

Solenoid valves (normally closed)

Hydrogen peroxide bottle pressure regulator

Engine controls regulator

Helium storage bottle

Solenoid valve (normally closed)

T - 4 sec disconnects

CD-9843-31

Engine controls
   Regulator pressure (gage), 303 to 327 N/cm^2 (440 to 475 psi)
   Relief valve pressure (gage), 327 to 362 N/cm^2 (475 to 525 psi)
Hydrogen peroxide bottle
   Pressure regulator (gage), 205 to 217 N/cm^2 (297 to 315 psi)
   Relief valve pressure (gage), 221 to 244 N/cm^2 (320 to 354 psi)
Orifice-tank pressurization
   Oxygen tank, 0.109-cm (0.043-in.) diameter
   Hydrogen tank, 0.226-cm (0.089-in.) diameter
Constant bleed orifice, 2211 standard cm^3/min (135 standard cu in./min)
Pressure switch (oxygen tank), 26 to 28 N/cm^2 abs (38 to 40 psia)
Helium bottle
   Volume, 76 167 cm^3 (4650 cu in.)
   Initial pressure, 1913 N/cm^2 abs (2775 psia)

(a) Tank pressurization and propulsion pneumatics subsystems.

Figure VI-23. - Centaur pneumatics system, AC-16.

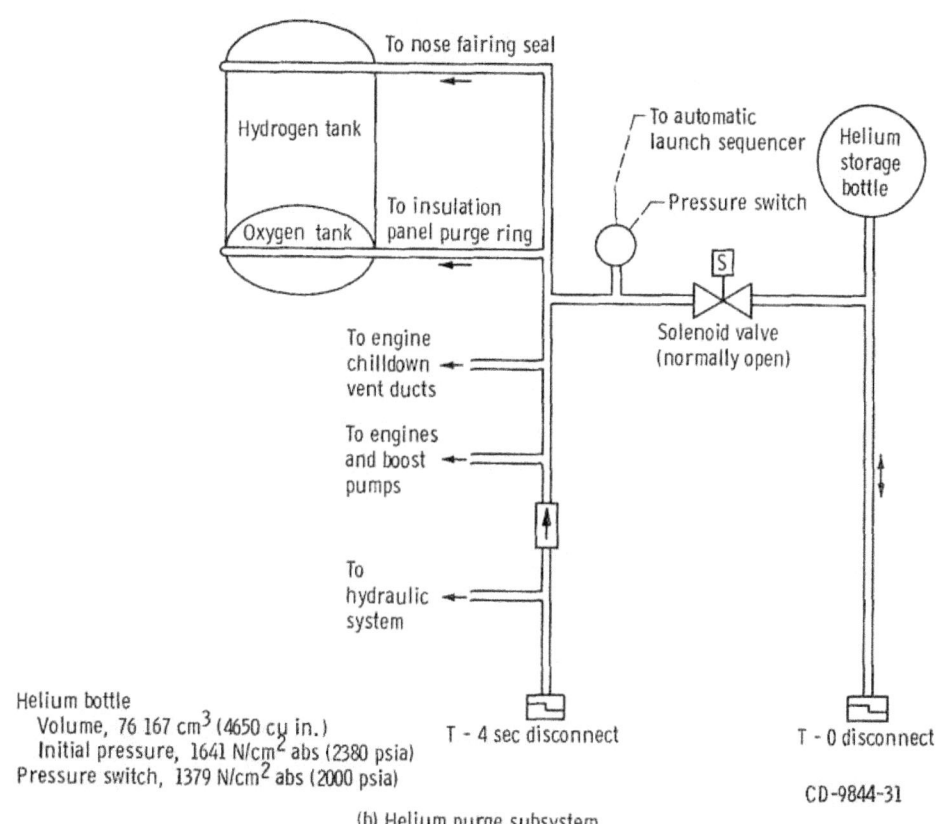

Helium bottle
    Volume, 76 167 cm^3 (4650 cu in.)
    Initial pressure, 1641 N/cm^2 abs (2380 psia)
Pressure switch, 1379 N/cm^2 abs (2000 psia)

(b) Helium purge subsystem.

Figure VI-23. - Concluded.

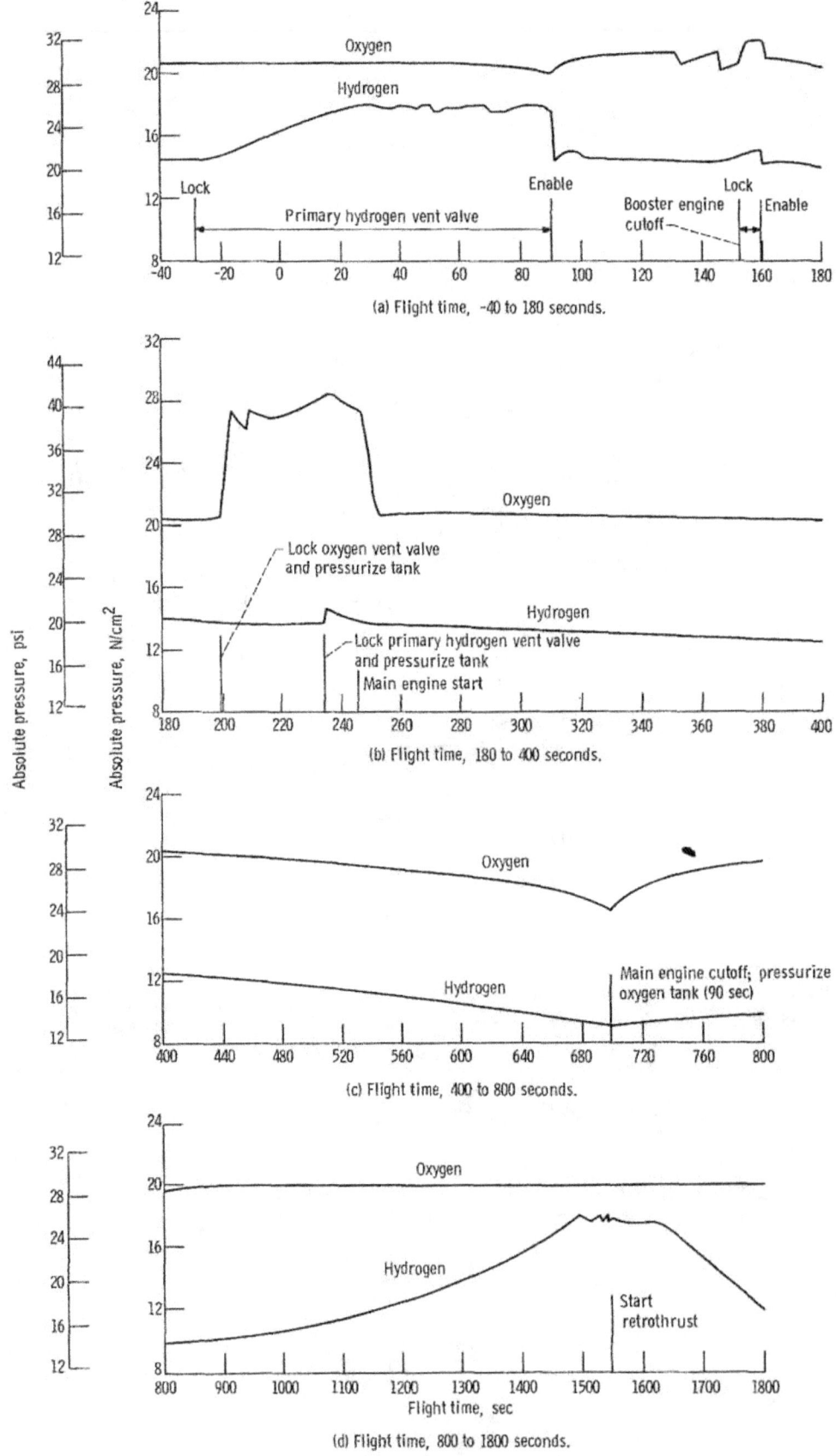

Absolute pressure, psi

Absolute pressure, N/cm²

(a) Flight time, −40 to 180 seconds.

(b) Flight time, 180 to 400 seconds.

(c) Flight time, 400 to 800 seconds.

(d) Flight time, 800 to 1800 seconds.

Figure VI-24. - Centaur tank pressure history, AC-16.

HYDRAULIC SYSTEMS

by Eugene J. Cieslewicz and Eugene J. Fourney

Atlas

System description. - Two hydraulic systems (figs. VI-25 and VI-26) are used on the Atlas vehicle to supply fluid power for operation of sustainer engine control valves and for thrust vector control of all engines. One system is used for the booster engines and the other for the sustainer engine.

The booster hydraulic system provides power solely for gimbaling the two thrust chambers. System pressure is supplied by a single, pressure-compensated, variable-displacement pump driven by the engine turbopump accessory drive. Additional components of the system include four servocylinders, a high-pressure relief valve, an accumulator, and a reservoir. Engine gimbaling in response to flight control commands is accomplished by the servocylinders, which provide separate pitch, yaw, and roll control during the booster engine phase of flight. The maximum booster engine gimbal angle capability is $\pm 5^O$ in the pitch and yaw planes.

The sustainer stage uses a system similar to that of the booster but, in addition, provides hydraulic power for sustainer engine control valves for gimbaling of the two vernier engines. Vehicle roll control is accomplished during the sustainer phase by differential gimbaling of the vernier engines. Actuator limit travel for the vernier engines is $\pm 70^O$ and for the sustainer engine is $\pm 3^O$.

System performance. - Hydraulic system pressure data for both the booster and sustainer circuits was normal. Pressures were stable throughout the boost flight phase. The transfer of fluid power from ground to airborne hydraulics systems was normal. Pump discharge absolute pressures increased from 1290 newtons per square centimeter (1870 psi) at T - 2 seconds to flight levels of 2100 newtons per square centimeter (3050 psi) in less than 2 seconds. Starting transients produced a normal overshoot of about 10 percent in the pump discharge pressure. Absolute pressure in the sustainer hydraulic and booster circuits stabilized at 2100 and 2171 newtons per square centimeter (3050 and 3150 psi), respectively.

During the booster phase of flight the sustainer pump return pressure transducer measurement (AH601P) indicated return pressure transients. Some transients as high as 295 newtons per square centimeter (430 psi) gage were recorded. Normal sustainer pump return pressure is approximately 44.8 newtons per square centimeter (65 psi) gage. This condition has been noted on most SLV-3C Atlas boosters. The cause of this anomaly is unknown at this time; however, it had no adverse effects on the performance of the hydraulic system.

Centaur

System description. - Two separate but identical hydraulic systems (fig. VI-27) are used on the Centaur stage. Each system gimbals one engine for pitch, yaw, and roll control. Each system consists of two servocylinders and a power package coupled to the engine. The power package contains high- and low-pressure pumps, reservoir, accumulator, pressure-intensifying bootstrap piston, and relief valves for pressure regulation. High-pressure power is provided by a constant-displacement vane-type pump driven by the liquid-oxygen turbopump accessory drive shaft. An electrically powered recirculation pump is used to provide low pressure for engine gimbaling requirements during prelaunch checkout, and during flight to aline the engines prior to main engine start. It is also used for limited thrust vector control during the propellant tank discharge for the Centaur retrothrust operation. Maximum engine gimbal capability is $\pm 3^{\circ}$.

System performance. - The hydraulic system properly performed all guidance and flight control commands throughout the flight. System pressures and temperatures as a function of flight time are shown in figures VI-28 and VI-29.

Activation of the low-pressure recirculation pumps provided absolute hydraulic pressures of 71.8 newtons per square centimeter (104.1 psi) for the C-1 engine and 70.3 newtons per square centimeter (102.0 psi) for the C-2 engine system. These pumps provided pressure and flow for centering the engines prior to main engine start. Main system absolute pressure in the C-1 and C-2 systems reached 750.6 and 754.2 newtons per square centimeter (1088.6 and 1093.8 psi), respectively, at main engine start. Manifold temperatures rose from 282.9 and 283.5 K (49.8° and 50.9° F), respectively, for C-1 and C-2 at main engine start to 347.1 and 349.0 K (165.4° and 168.8° F) at main engine cutoff. After cutoff the temperatures slowly decreased to steady values of 334.3 K (142.3° F) on C-1 and 332.9 K (139.8° F) on C-2.

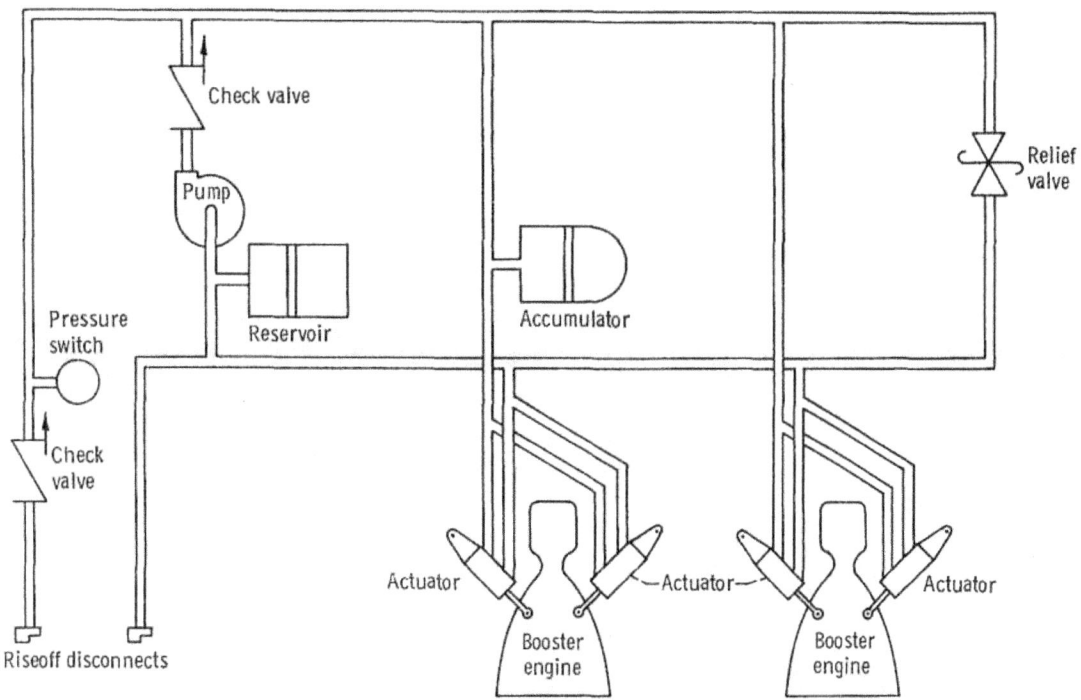

Figure V-25. - Atlas booster hydraulic system, AC-16.

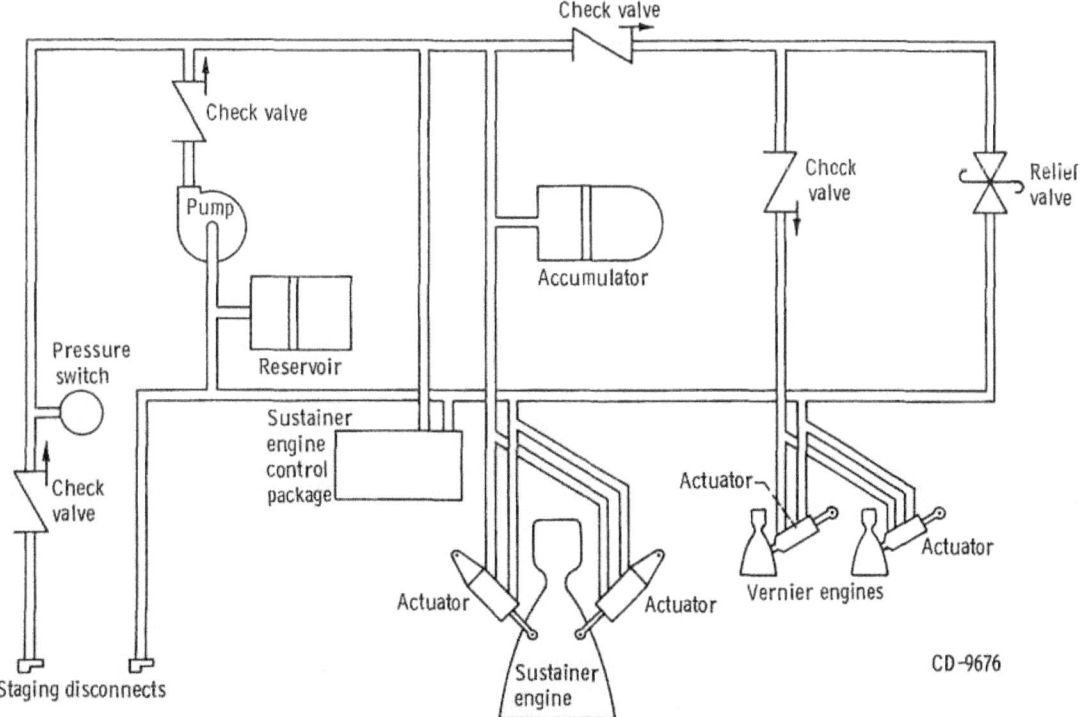

Figure VI-26. - Atlas sustainer hydraulic system, AC-16.

CD-9676

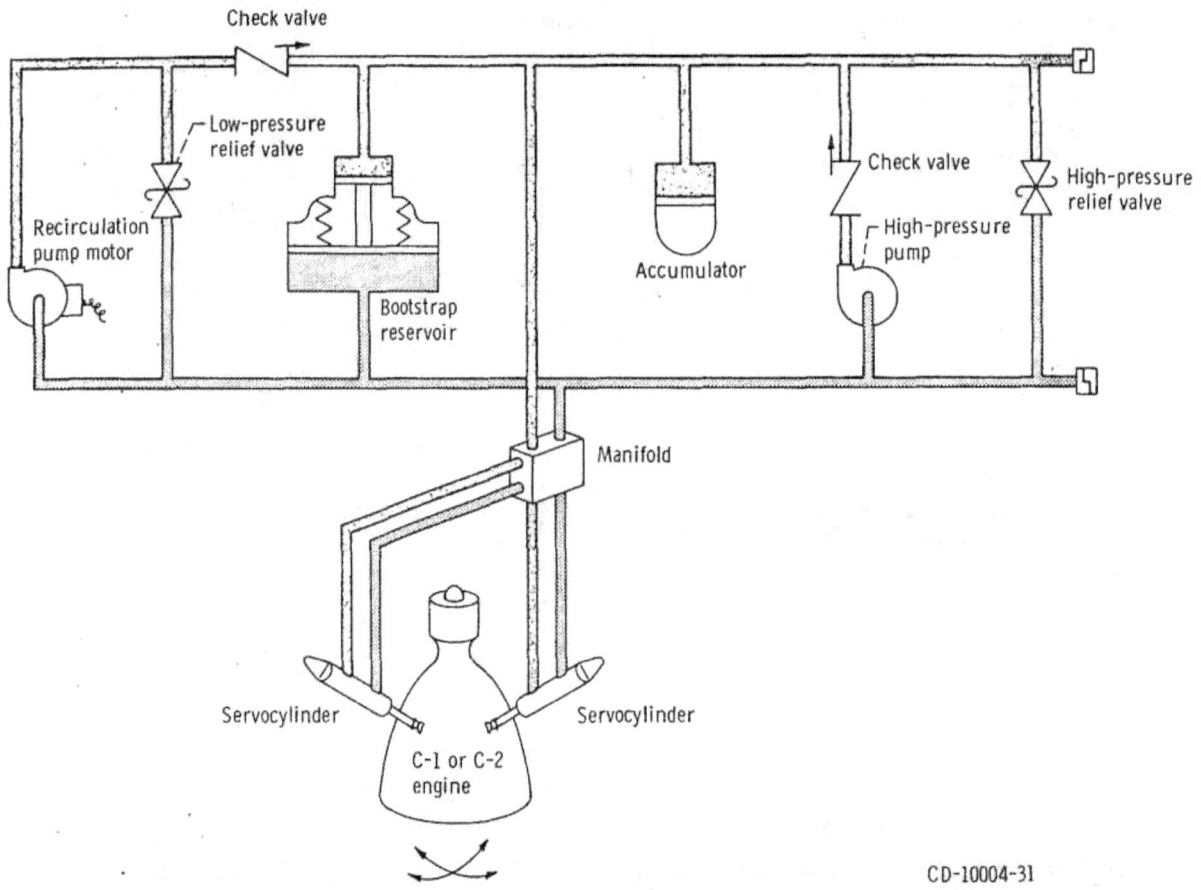

Figure VI-27. - Centaur hydrualic system, AC-16.

CD-10004-31

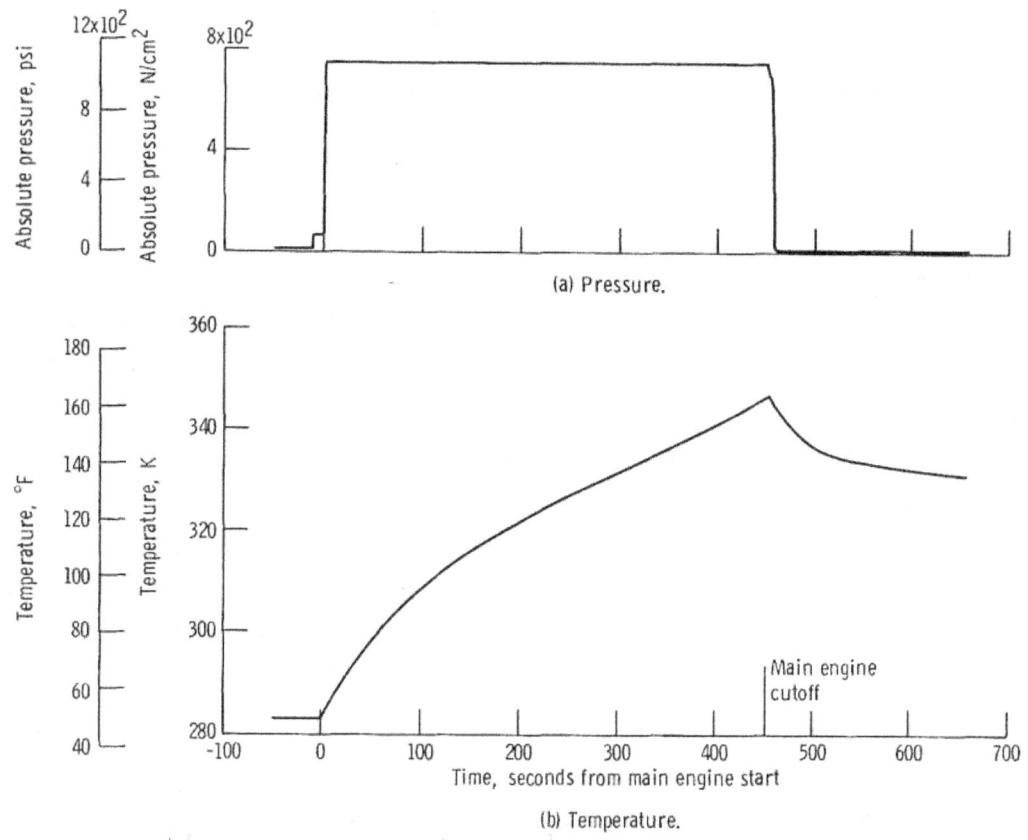

(a) Pressure.

(b) Temperature.

Figure VI-28. - Hydraulic system pressure and temperature for C-1 engine, AC-16.

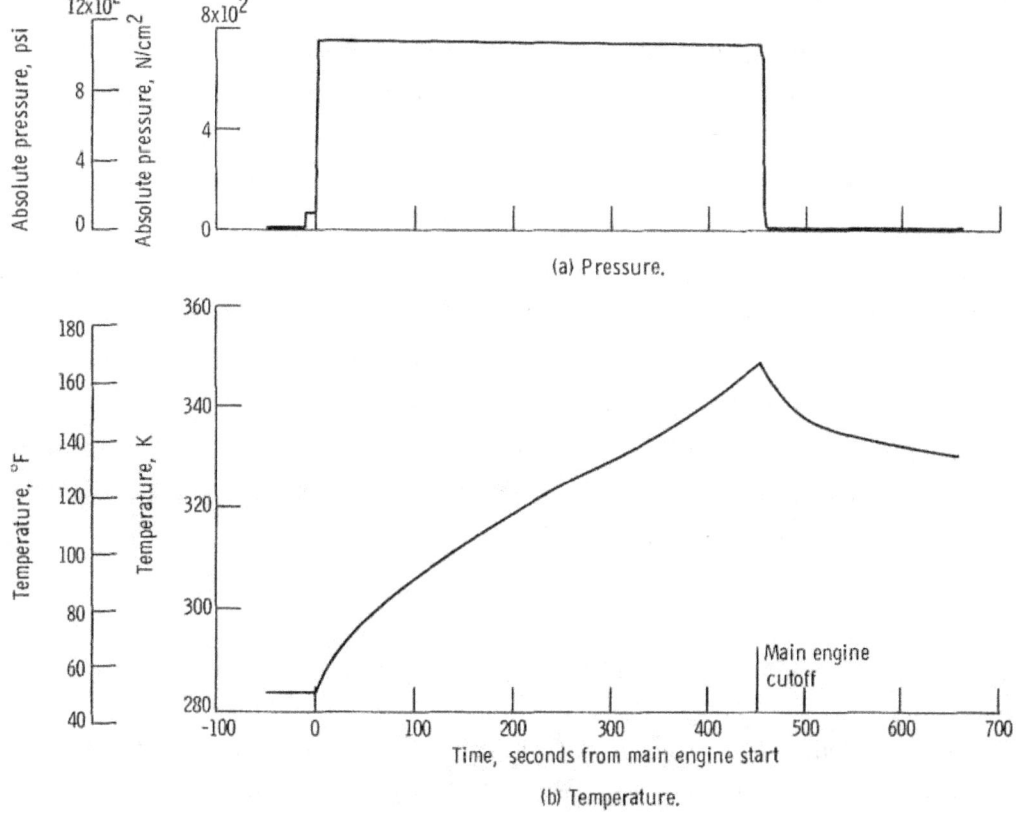

(a) Pressure.

(b) Temperature.

Figure VI-29. - Hydraulic system pressure and temperature for C-2 engine, AC-16.

VEHICLE STRUCTURES

by James F. Harrington, Robert C. Edwards, and Dana H. Benjamin

Atlas Structures

System description. - The primary Atlas vehicle structure is provided by the propellant tanks. These tanks are thin-walled, pressure-stabilized, monocoque sections of welded stainless-steel construction (fig. VI-30). They require internal pressure in order to maintain structural stability. The tensile strength of the tank material determines the maximum allowable pressure in the propellant tanks.

The maximum allowable and minimum required tank pressures presented in figure VI-31 are computed by using maximum design loads (as opposed to actual flight loads) with appropriate factors of safety. These required tank pressures are not constant because of varying aerodynamic loads, inertial loads, and ambient pressure during flight.

The Atlas vehicle is subjected to its highest design bending load between $T + 40$ and $T + 100$ seconds. The bending, inertia, and aerodynamic drag create compressive loads in the fuel and oxidizer tank skin. These loads are resisted by internal pressure to prevent buckling of the skin.

The maximum allowable differential pressure between the oxidizer and fuel tanks is limited by the strength of the Atlas intermediate bulkhead. The fuel tank pressure must always be greater than the oxidizer tank pressure to stabilize the intermediate bulkhead (prevent bulkhead reversal).

System performance. - The Atlas oxidizer and fuel tank ullage pressures did not approach the maximum allowable pressure during flight. The oxidizer and fuel tank ullage pressures were greater than the minimum required to resist the combined bending and axial design loads between $T + 40$ and $T + 100$ seconds (fig. VI-31). The bulkhead differential pressure was within the maximum allowable and minimum required pressure limits for all periods of flight (fig. VI-31).

The increase of longitudinal inertia force was as expected. A maximum value of 5.74 g's was reached at booster engine cutoff $(T + 152.1$ sec). This value was within the specified limits of 5.587 to 5.813 g's.

Centaur Structures

System description. - The primary Centaur vehicle structure is provided by the propellant tanks. These tanks are thin-walled, pressure-stabilized, monocoque sections of welded stainless-steel construction (fig. VI-32). They require internal pressure in order to maintain structural stability. The tensile strength of the tank material determines the maximum allowable pressure in the propellant tanks.

The maximum allowable and minimum required tank pressures presented in figures VI-33(a) to (c) are computed by using maximum design loads (as opposed to actual flight loads) with appropriate factors of safety. The maximum allowable and minimum required tank pressures are not constant because of varying loads and varying ambient pressure during flight. The tank locations and criteria which determine the maximum allowable and minimum required tank pressures during different phases of flight are described in figure VI-34.

The oxidizer tank pressure most closely approaches the maximum allowable pressure at booster engine cutoff (T + 152.1 sec), when the high inertial load causes maximum tension stresses on the aft bulkhead (fig. VI-33(a)). The minimum required oxidizer tank pressure for aft bulkhead stability is not pertinent because this required pressure is always less than the pressure required for intermediate bulkhead stability.

The strength of the fuel tank is governed by the capability of the conical section of the forward bulkhead to resist hoop stress. Thus, the differential pressure across the forward bulkhead determines the maximum allowable fuel tank pressure.

The minimum required fuel tank pressure was higher for AC-16 than for previous flights because of the heavier and longer nose fairing and payload. However, the only required adjustment to the fuel tank pressure from previous flights was an increase during the launch phase from T + 0 to T + 5 seconds (fig. VI-33(a)).

The margin between fuel tank ullage pressure and the minimum required pressure was least during the following events:

(1) Prior to launch, the payload and nose fairing impose compression loads on the cylindrical skin at station 409.6 due to gravity and ground winds.

(2) During the launch phase from T + 0 to T + 5 seconds, the payload and nose fairing impose compression loads on the cylindrical skin at station 409.6 due to longitudinal and lateral inertia and vibration.

(3) From T + 40 to T + 100 seconds, the Centaur was subjected to maximum design bending moments. The combined loads due to inertia, aerodynamic drag, and bending impose compression on the cylindrical skin at station 409.6.

The maximum allowable differential pressure between the oxidizer and fuel tanks was limited by the strength of the Centaur intermediate bulkhead. The maximum design allowable differential pressure was 15.9 newtons per square centimeter (23.0 psi). The oxidizer tank pressure must always be greater than the combined fuel tank pressure and hydrostatic pressure of the hydrogen fuel; this is necessary to stabilize the intermediate bulkhead (prevent bulkhead reversal).

System performance. - The Centaur fuel and oxidizer tank ullage pressure profiles are compared with the design limits in figures VI-33(a) to (c).

The oxidizer tank pressure was less than the maximum allowable at booster engine cutoff (T + 152.1 sec) and all other periods of flight. The oxidizer tank pressure was maintained above the minimum required for aft bulkhead stability during all periods of flight. At no time during the flight did the fuel tank ullage pressure exceed the maximum allowable pressure. The fuel tank ullage pressure was safely above the minimum required pressure at all times. The differential pressure was less than the maximum allowable 15.9 newtons per square centimeter (23.0 psi) for all periods of flight. The oxidizer tank pressure was always greater than the combined fuel tank ullage pressure and hydrostatic pressure of hydrogen fuel.

Vehicle bending loads. - Flight bending loads were determined at station 125 on the Centaur fixed fairing. The purpose of this measurement was to ascertain the effect of the longer nose fairing on vehicle bending loads; no unexpected bending loads were measured. The maximum bending moment during flight was 16.0×10^{6} centimeter-newtons (1.4×10^{6} in.-lbf) at T + 70 seconds (fig. VI-35). This was well within the design limit bending moment of 32.0×10^{6} centimeter-newtons (2.8×10^{6} in.-lbf).

Vehicle Dynamic Loads

The Atlas-Centaur launch vehicle receives dynamic loading from three major sources: (1) external loads from aerodynamic and acoustic sources; (2) transients from engines starting and stopping and from the separation systems; and (3) loads due to dynamic coupling between major systems.

Research and development flights of the Atlas-Centaur have shown that these loads are within the structural limits established by ground test and model analysis. For operational flights such as AC-16, the number of dynamic flight measurements is limited by telemetry capacity. The instruments used and the parameters measured are given in the following table:

Instruments	Corresponding parameters
Low-frequency-range accelerometer	Vehicle longitudinal vibration
Centaur pitch rate gyro	Vehicle pitch plane vibration
Centaur yaw rate gyro	Vehicle yaw plane vibration
Angle-of-attack sensor	Vehicle aerodynamic loads
High-frequency-range accelerometer	Local spacecraft vibration

Launch vehicle longitudinal vibrations measured on the Centaur forward bulkhead are presented in figure VI-36, which depicts the presence of specific responses at the times noted. The frequency and amplitude of the vibration data measured on this flight are shown together with similar data from other flights.

During launcher release, longitudinal vibrations were excited. The amplitude and frequency of these vibrations were similar to those observed on vehicles prior to AC-16. Atlas intermediate bulkhead pressure fluctuations were the most significant effects produced by the launcher-induced longitudinal vibrations. The peak pressure fluctuations computed from these vibrations were 1.2 newtons per square centimeter (1.7 psi). Since the minimum bulkhead differential pressure measured during this time was 8.9 newtons per square centimeter (13 psi) (fig. VI-31), the calculated minimum differential pressure was 7.7 newtons per square centimeter (11 psi). The minimum design allowable differential pressure across the bulkhead is 1.4 newtons per square centimeter (2.0 psi).

During Atlas flight between T + 82 and T + 154 seconds, intermittent longitudinal vibrations of 0.10 g, 12 hertz were observed on the forward bulkhead. These vibrations are believed to be caused by dynamic coupling between structure, engines, and propellant lines (commonly referred to as POGO). The AC-16 vehicle was flown with a longer cylindrical section added to the nose fairing. As a result, AC-16 vehicle length was 17.9 feet (5.45 m) longer than AC-13, AC-14, and AC-15 vehicles and 22.2 feet (6.77 m) longer than AC-10, AC-12, and AC-11 vehicles. As a result, the parameters controlling the frequency and amplitude of these vibrations were changed slightly, but not significantly. For a detailed discussion of this low-frequency longitudinal vibration see reference 1.

During the booster engine thrust decay, short-duration longitudinal vibrations of 0.6 g, 12 hertz were observed. The analytical models did not indicate significant structural loading due to these transients.

During the booster phase of flight, the vehicle vibrates in the pitch plane and the yaw plane as an integral body at all of its natural frequencies. Previous analyses and tests have defined these natural frequencies or modes and the shapes which the vehicle assumes when the modes are excited. The rate gyros on the Centaur provide data for determining the deflection of these modes. The maximum first-mode deflection was seen in the pitch plane at T + 132 seconds (fig. VI-37). The deflection was less than 4 percent of the allowable deflection. The maximum second-mode deflection was seen in the pitch plane at T + 42 seconds (fig. VI-38). The deflection was less than 13 percent of the allowable deflection.

Predicted angles of attack were based upon upper-wind data obtained from weather balloons released before the time of launch. Vehicle bending moments were calculated by using predicted angles of attack, booster engine gimbal angle data, vehicle weights, and vehicle stiffnesses. These moments were added to axial load equivalent moments and to moments resulting from random dispersions. The most significant dispersions considered were uncertainties in launch vehicle performance, vehicle center-of-gravity offset, and upper-atmosphere wind.

The total equivalent predicted bending moment (based upon wind data) was divided by the design bending moment allowable to obtain the predicted structural capability ratio shown in figure VI-39. This ratio is expected to be greatest between $T + 60$ and $T + 88$ seconds because of the high aerodynamic loads during this period. The maximum structural capability ratio predicted for this period was 0.87.

Transducers located on the nose fairing cap provided differential pressure measurements in the pitch plane and the yaw plane. Total pressure was computed from a trajectory reconstruction. Predicted angles of attack are shown in figures VI-40 and VI-41. The angles of attack derived from in-flight pressure data were not available for the total flight at the time of publication. However, a spot check showed that the predicted and actual angles of attack were within the expected dispersion during the most critical flight times. It follows, therefore, that the maximum predicted capability ratio of 0.87 was not exceeded.

Local shock and vibration were measured continuously by three piezoelectric accelerometers in the spacecraft area. The accelerometers were located on the forward end of the payload adapter (Centaur station 68). These accelerometers together with their amplifiers had a frequency response of 2100 hertz, and in all cases the telemetry channel Interrange Instrumentation Group (IRIG) filter frequency was less than this value. Therefore, the frequency range over which one could expect to obtain unattenuated data was limited by the standard IRIG filter frequency for that channel. In addition to the three high-frequency accelerometers, there were six low-frequency accelerometers on the payload adapter (Centaur station 85) located on the x and y axes. These accelerometers were sensitive in the tangential and longitudinal directions.

A summary of the most significant shock and vibrations levels measured continuously by the three high-frequency and six low-frequency accelerometers on AC-16 together with similar data from AC-10, AC-12, AC-11, AC-13, AC-14, and AC-15 are shown in table VI-VIII. The major reason for the apparent discrepancy between the vibration data obtained from the AC-16 vehicle and that obtained from the AC-10, AC-12, and AC-11 vehicles is that the locations of the three high-frequency vibration accelerometers on the two vehicle configurations were different. On AC-10, AC-12, and AC-11, the one continuous accelerometer was installed on one of the retromotor attach points of the Surveyor spacecraft; whereas on AC-13, AC-14, AC-15, and AC-16 the high-frequency accelerometers were installed on the forward end of the payload adapter. From the data it can be seen that the payload adapter receives more vibration (especially during launch transient) than the spacecraft structure. The steady-state vibration levels were highest near lift-off, as expected. The maximum level of the shock loads (24 g's) on AC-16 occurred at nose fairing jettison. These shock levels were of short duration (~0.025 sec) and did not provide significant loads. An analysis of the data indicates that the levels were well within spacecraft qualification levels.

TABLE VI-VIII. - COMPARISON OF MAXIMUM SHOCK AND VIBRATION LEVELS AT MARK EVENTS[a]

(a) Previous Atlas-Centaur flights

Flight event	Flight								
	AC-10	AC-12	AC-11	AC-13	AC-14	AC-15	AC-13	AC-14	AC-15
	Accelerometer location								
	Retromotor attachment 1, station 125; quadrant I-IV; longitudinally sensitive; analysis band, 10 to 790 Hz			Payload adapter, station 129; quadrant III; longitudinally sensitive; analysis band, 10 to 790 Hz			Payload adapter, station 129; quadrant I-IV; radially sensitive[b]		

Launch:
  Acceleration, g's (rms)
  Frequency, Hz
	0.68
	165
		0.65
		165
			0.53
			160 to 170
				1.5
				150, 300
					1.4
					163, 296, 397
						1.2
						225, 296, 389, 425
							2.5
							470
								2.9
								485
									2.1
									462, 478, 525

Booster engine cutoff:
  Acceleration, g's
  Frequency, Hz
	0.8
	11
		1.2
		17
			0.7
			12
				1.2
				13
					0.96
					14
						1.7
						13
							0.83
							4.5
								0.38
								4.7
									0.6
									4.5

Booster jettison:
  Acceleration, g's
  Frequency, Hz
	0.5
	16
		0.46
		16
			0.3
			23
				<1/2
				---------- 	
					<1/2
					----------- 	
						<1/2
						14
							<1/2
							--------- 	
								<1/2
								--------- 	
									<1/2
									6

Insulation panel jettison:
  Acceleration, g's
  Frequency, Hz
	10
	700
		10.1
		600 to 700
			12
			600 to 700
				~14
				500 to 600
					~13
					500 to 600
						~14
						500 to 600
							~12
							500 to 600
								~12
								500 to 600
									~9
									500 to 600

Nose fairing jettison:
  Acceleration, g's
  Frequency, Hz
	1.4
	1.4
	32
		0.49
		0.49
		20
			1.1
			1.1
			32
				2.1
				2.1
				400 to 500
					2.7
					2.7
					400 to 500
						3.2
						3.2
						400 to 500
							2.0
							2.0
							400 to 500
								1.6
								1.6
								400 to 500
									2.0
									2.0
									400 to 500

Atlas-Centaur separation:
  Acceleration, g's
  Frequency, Hz
	12
	600
		13
		600 to 700
			12
			700
				~14
				500 to 600
					~13
					500 to 600
						~14
						500 to 600
							~12
							500 to 600
								~12
								500 to 600
									~9.5
									500 to 600

Main engine first start:
  Acceleration, g's
  Frequency, Hz
	0.38
	20
		0.5
		20 to 21
			0.4
			19 to 20
				0.5
				22
					0.6
					20
						0.5
						22
							(c)
							(c)
								(c)
								(c)
									(c)
									(c)

Main engine first cutoff:
  Acceleration, g's
  Frequency, Hz
	1.14
	33
		0.95
		22
			2.0
			27
				1.6
				23
					0.9
					23
						0.9
						23
							0.9
							400 to 500
								0.7
								400 to 500
									0.7
									400 to 500

Main engine second start:
  Acceleration, g's
  Frequency, Hz
	(d)
	(d)
		0.66
		20 to 22
			(d)
			(d)
				0.7
				20 to 30
					0.6
					30
						0.6
						23
							(c)
							(c)
								0.5
								400 to 500
									(c)
									(c)

Main engine second cutoff:
  Acceleration, g's
  Frequency, Hz
	(d)
	(d)
		0.97
		24
			(d)
			(d)
				0.9
				30
					0.9
					30
						0.9
						30
							1.1
							480
								1.5
								480
									3
									400 to 500

[a]Maximum shock and vibration levels at mark events are given in terms of maximum single amplitude (in g's) and the most predominant frequency (in Hz) except for rms levels which represent maximum levels observed at launch.

[b]A nonstandard Interrange Instrumentation Group filter of 600 Hz was used to analyze launch data on AC-13, AC-14, and AC-15. All other postlaunch data were analyzed with a 330-Hz filter.

[c]No detectable response.

[d]Single-burn missions.

TABLE VI-VIII. - Concluded. COMPARISON OF MAXIMUM SHOCK AND VIBRATION LEVELS AT MARK EVENTS

(b) AC-16 (OAO-II) mission

Flight event	Centaur station 68 (+x axis); frequency response, 2100 Hz (amplifier filter)			Centaur station 85					
	Analysis band, Hz			+y axis	+x axis	-y axis	+y axis	-y axis	-x axis
				Accelerator system response limited to -					
	10 to 330	10 to 790		100 Hz			70 Hz		
				Telemetry channel filter frequency limited to -					
				110 Hz	160 Hz	220 Hz	330 Hz	450 Hz	------------
	Accelerometer range, g's								
	±10	±20		-2.88 to 8.88	-3.00 to 9.00		-1.17 to 1.17	-1.16 to 1.17	-1.17 to 1.17
Launch: Acceleration, g's	2.5 (rms) 6.1	2.6 (rms) 6.5	2.6 (rms) 8.9	---------- 1.5	---------- 1.3	---------- 1.4	---------- 0.7	---------- 0.6	---------- 0.5
Frequency, Hz	310	280	280	100 to 600	100 to 600	100 to 600	100 to 600	100 to 600	100 to 600
Booster engine cutoff: Acceleration, g's	1.0	(c)	(c)	0.65	0.65	0.60	0.51	0.56	0.56
Frequency, Hz	8 to 9	(c)	(c)	12	12	12	90	90	90
Booster jettison acceleration[e], g's	0.5	(c)	(c)	0.2	0.2	0.2	0.36	0.25	0.22
Installation panel jettison: Acceleration, g's	11	24	24	1.7	1.5	1.3	0.43	0.35	0.45
Frequency, Hz	600 to 700	600 to 700	600 to 700	(e)	(e)	(e)	(e)	(e)	(e)
Sustainer engine cutoff: Acceleration, g's	(c)	(c)	(c)	0.1 / 0.8	0.1 / 0.5	0.1 / 0.8	0.16	0.14	0.36
Frequency, Hz	(c)	(c)	(c)	20 / 90	20 / 90	20 / 90	90	90	90
Atlas-Centaur separation acceleration[e], g's	5.5	12	19	1.1	1.2	0.8	0.3	0.3	0.45
Main engine start: Acceleration, g's	(c)	(c)	(c)	0.11	0.11	0.11	0.11	0.14	0.11
Frequency, Hz	(c)	(c)	(c)	21	21	21	(e)	(e)	(e)
Nose fairing jettison acceleration[e], g's	13	23	24	2.1	1.4	2.6	0.47	0.51	0.54
Main engine cutoff: Acceleration, g's	0.5	(c)	1.9	0.35	0.55	0.40	0.27	0.27	0.23
Frequency, Hz	320	(c)	(e)	(e)	(e)	(e)	(e)	(e)	(e)
Spacecraft separation: Acceleration, g's	12	(f)	(f)	2.6	1.4	2.1	1.0	1.0	0.5
Frequency, Hz	600 to 700	(f)	(f)	(e)	(e)	(e)	(e)	(e)	(e)

[c] No detectable response.

[e] Frequency response at amplitude noted was ≥100 Hz and <1000 Hz.

[f] Invalid data.

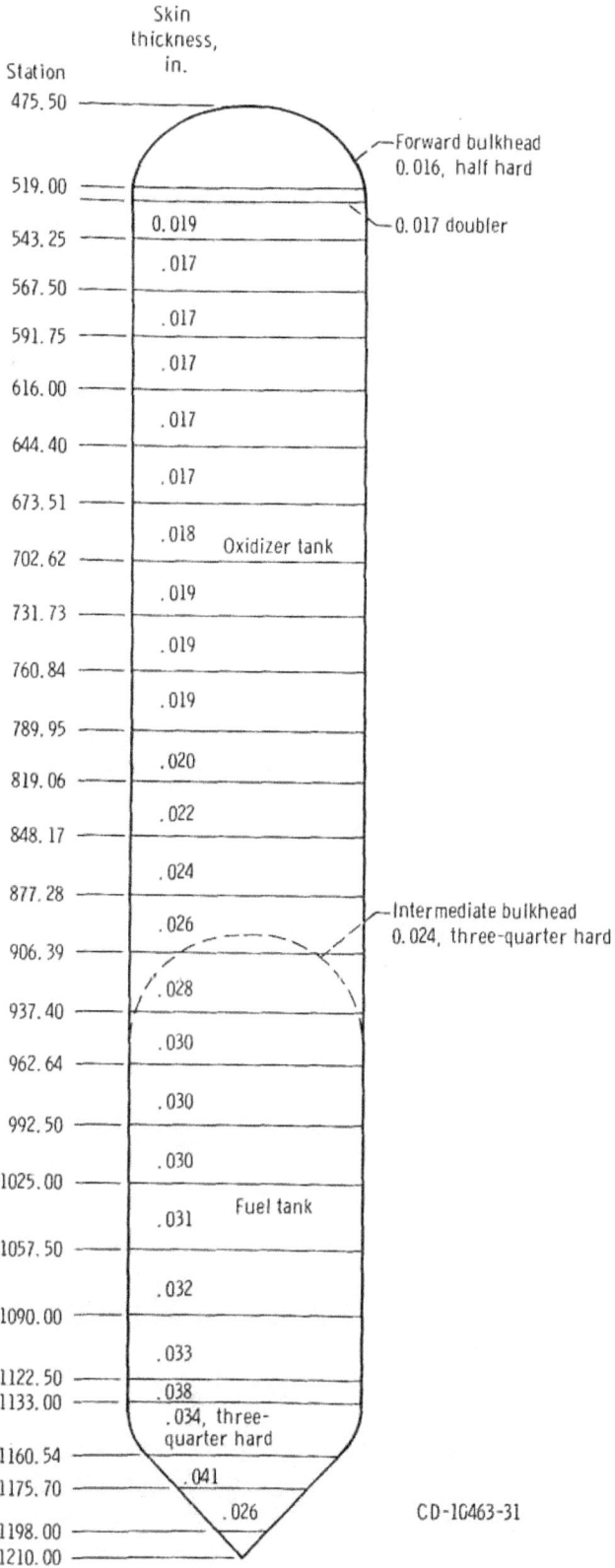

Figure VI-30. - Atlas propellant tanks, AC-16.

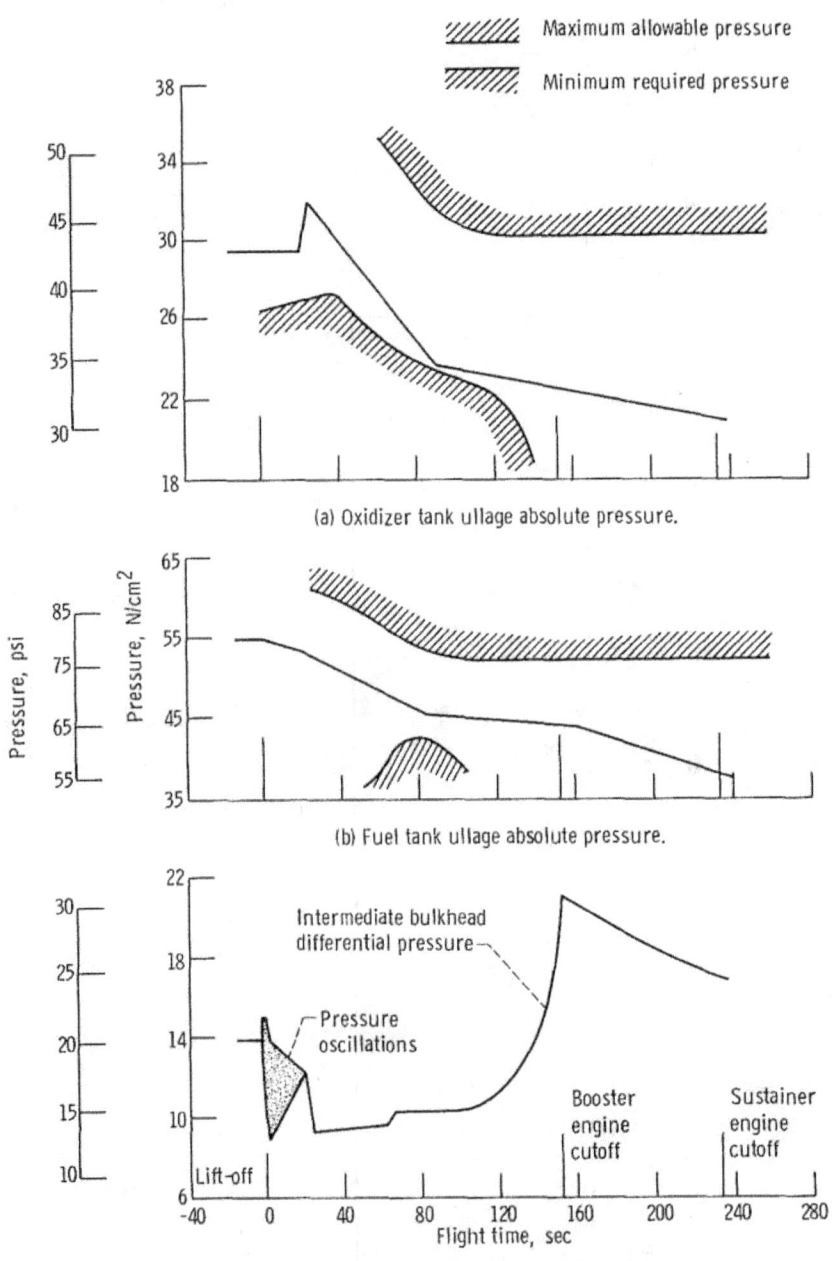

(a) Oxidizer tank ullage absolute pressure.

(b) Fuel tank ullage absolute pressure.

(c) Intermediate bulkhead differential pressure. Maximum allowable differential pressure, 42 newtons per square centimeter; minimum required differential pressure, 1.4 newtons per square centimeter.

Figure VI-31. – Atlas fuel and oxidizer tank pressures, AC-16.

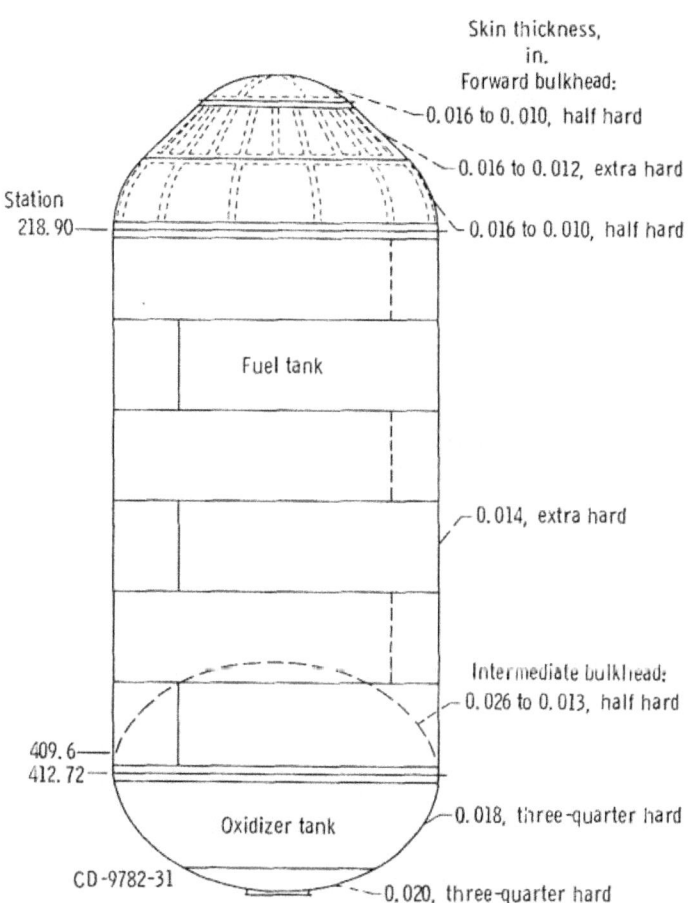

Figure VI-32. - Centaur propellant tanks, AC-16.
(All material 301 stainless steel, of hardness indicated.)

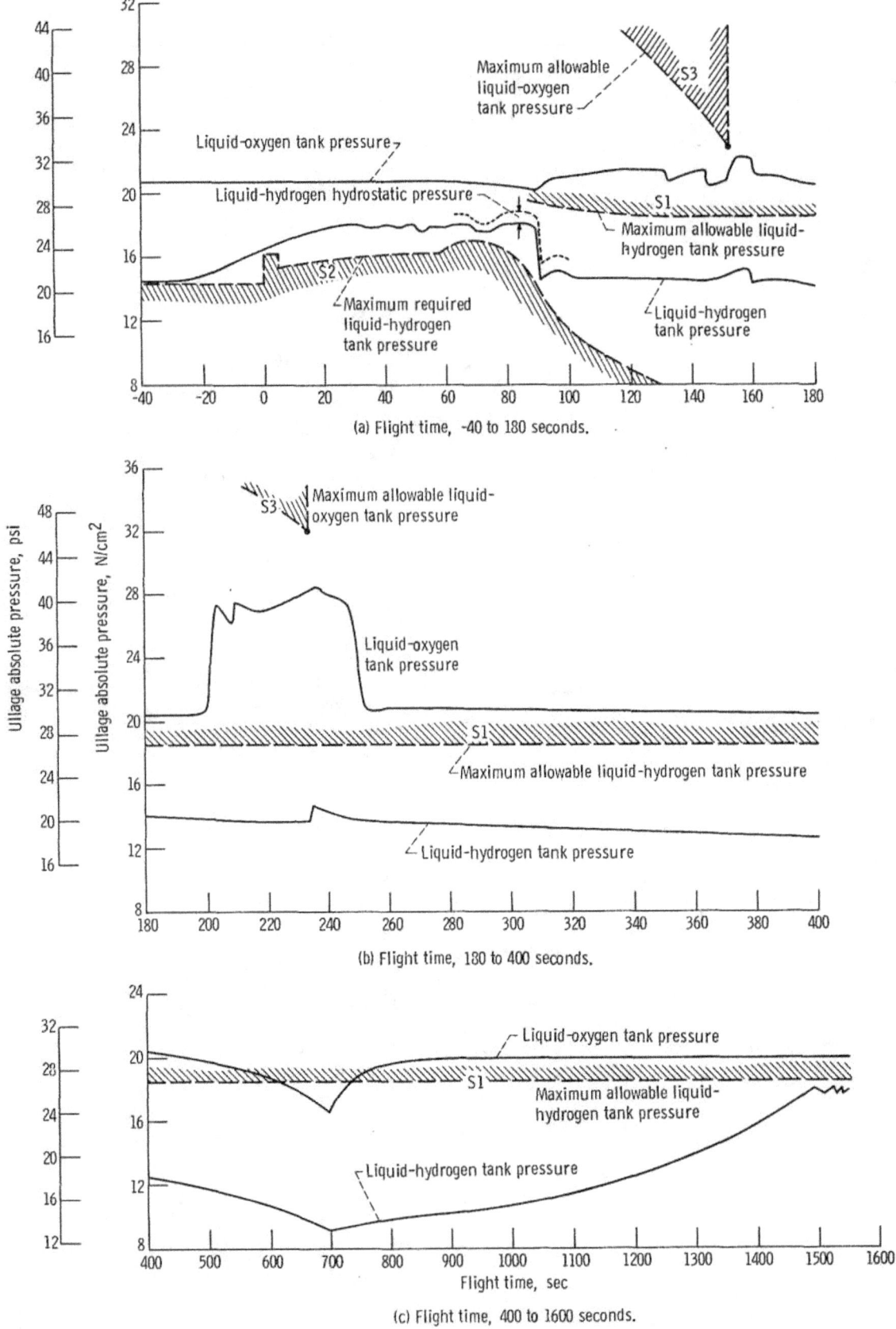

(a) Flight time, -40 to 180 seconds.

(b) Flight time, 180 to 400 seconds.

(c) Flight time, 400 to 1600 seconds.

Figure VI-33. - Centaur fuel and oxidizer tank pressure, AC-16. S1, S2, etc., indicate tank structure areas which determine allowable tank pressure (see fig. VI-34).

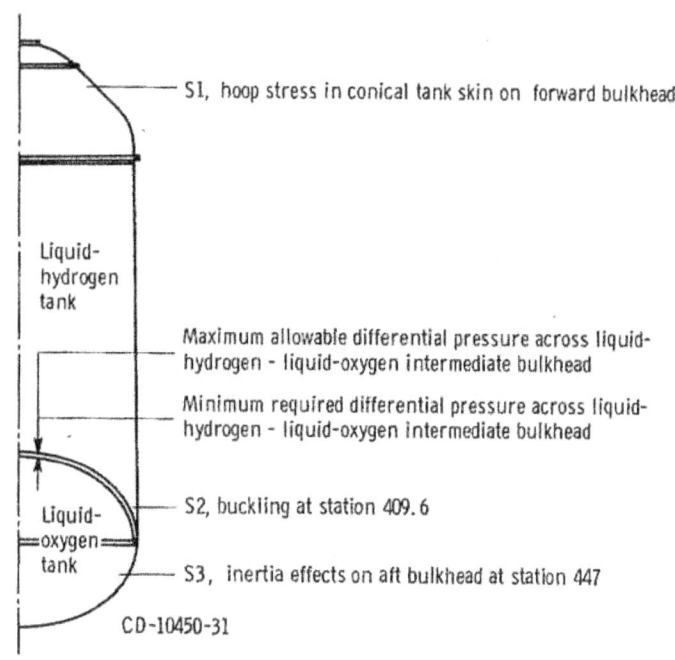

S1, hoop stress in conical tank skin on forward bulkhead

Liquid-hydrogen tank

Maximum allowable differential pressure across liquid-hydrogen - liquid-oxygen intermediate bulkhead

Minimum required differential pressure across liquid-hydrogen - liquid-oxygen intermediate bulkhead

Liquid-oxygen tank

S2, buckling at station 409.6

S3, inertia effects on aft bulkhead at station 447

CD-10450-31

Figure VI-34. - Tank locations and criteria which determine allowable pressures, AC-16.

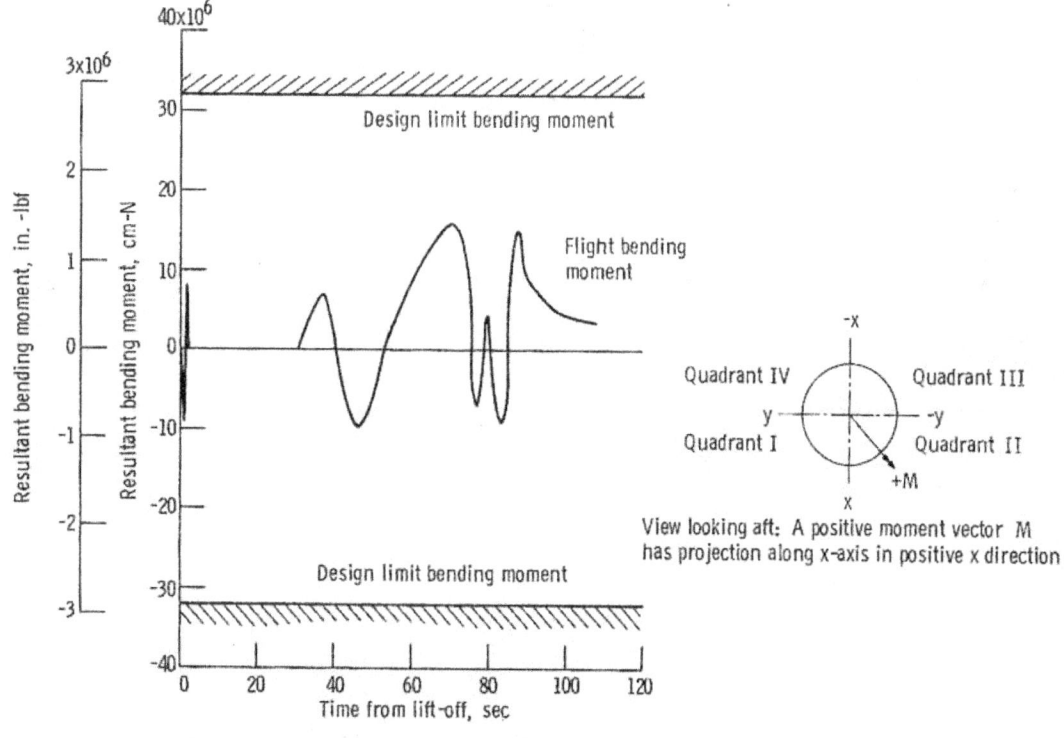

Figure VI-35. - Bending moment, fixed fairing at Centaur station 125, AC-16.

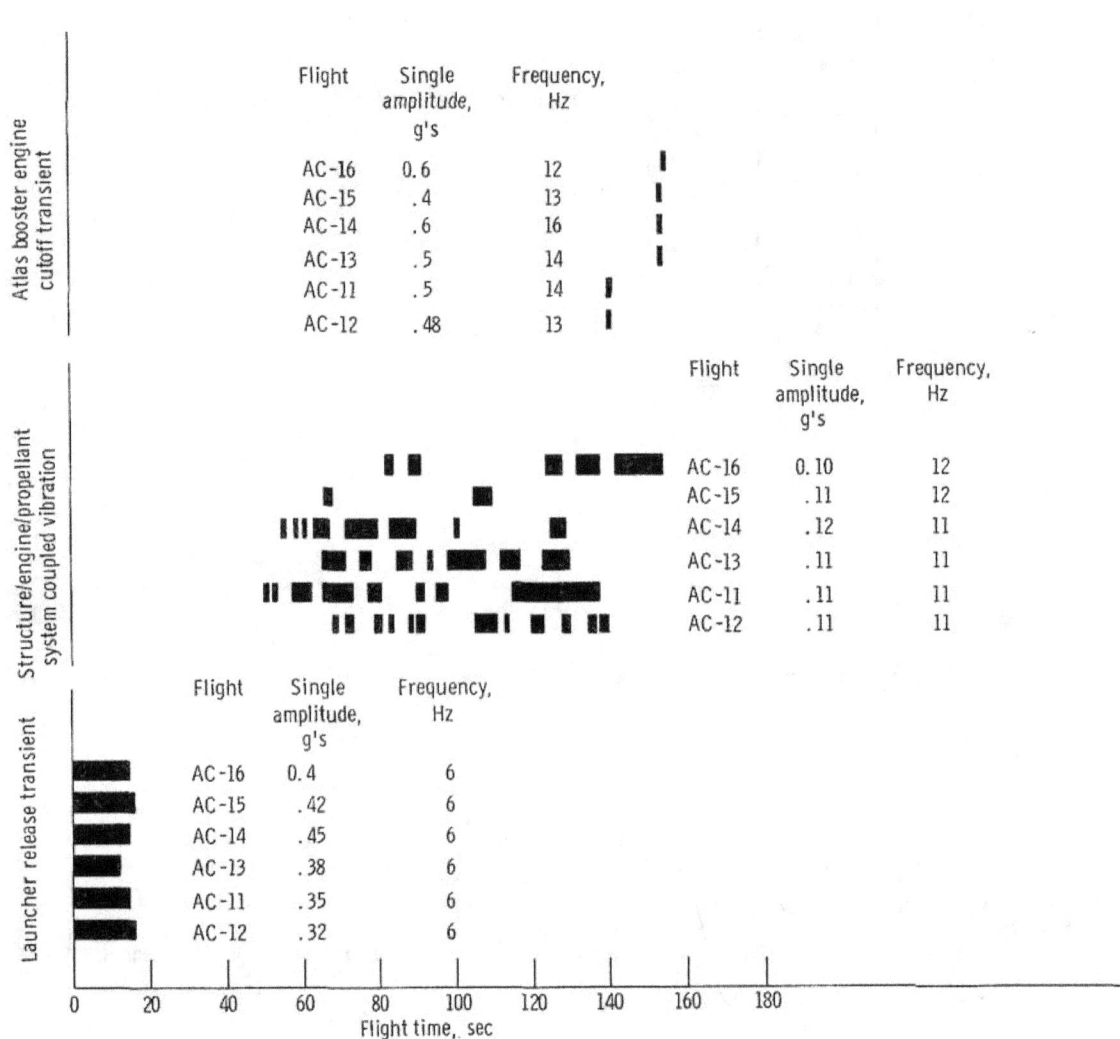

Figure VI-36. – Longitudinal vibrations for Atlas-Centaur flights.

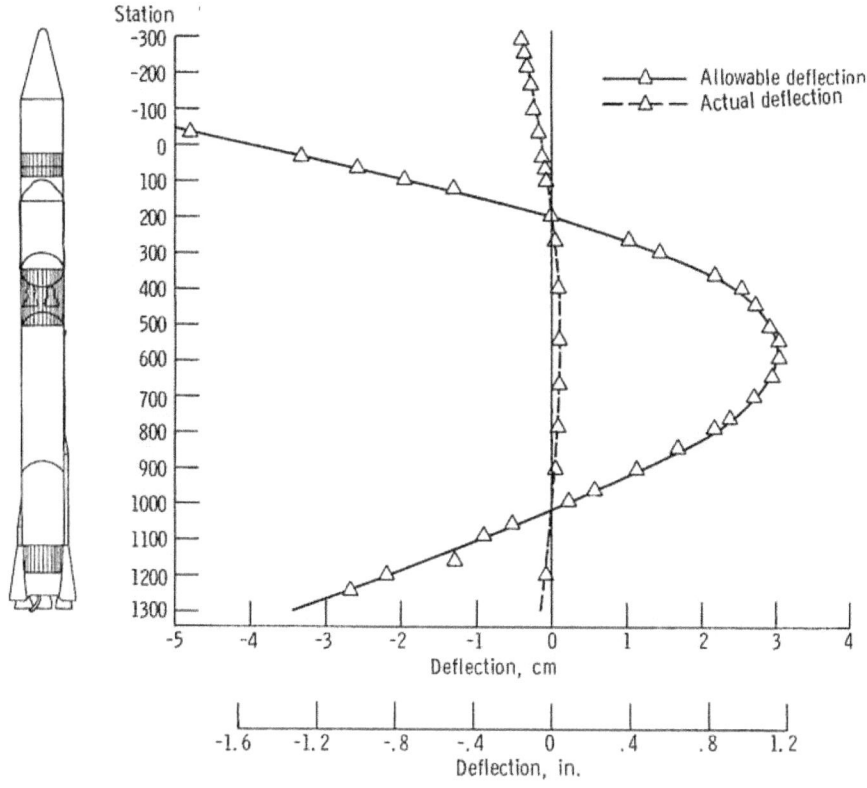

Figure VI-37. - Maximum pitch plane first-bending-mode amplitudes at
T + 132 seconds, AC-16.

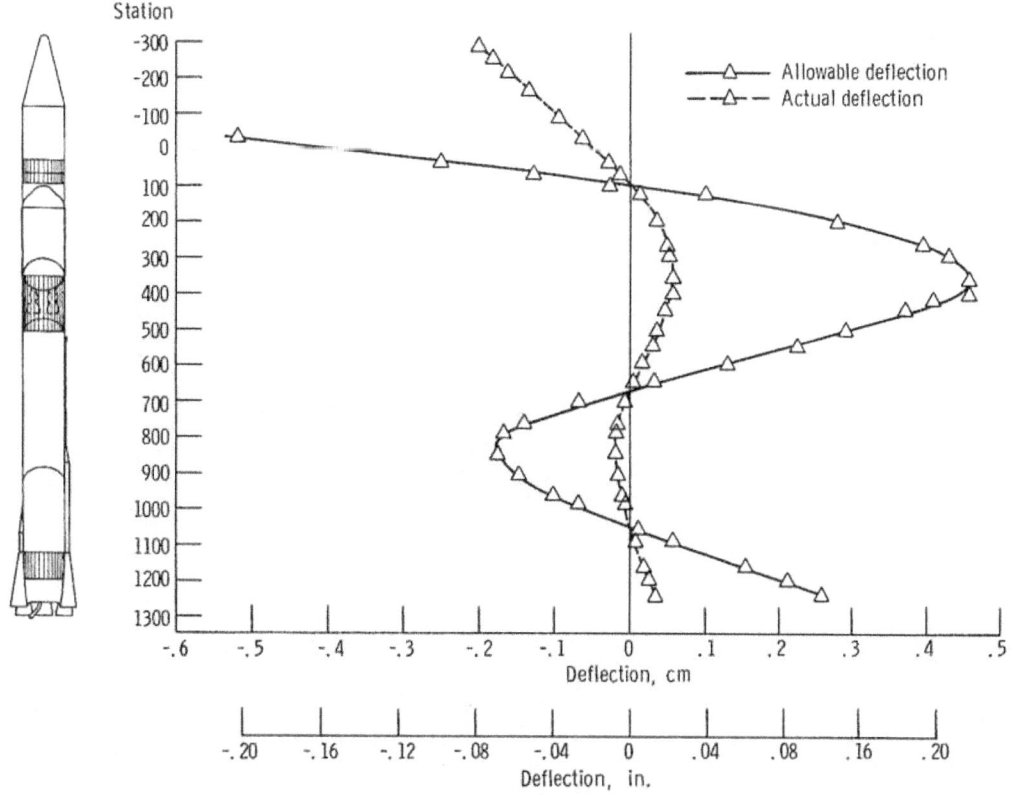

Figure VI-38. - Maximum pitch plane second-bending-mode amplitudes at T + 42 seconds,
AC-16.

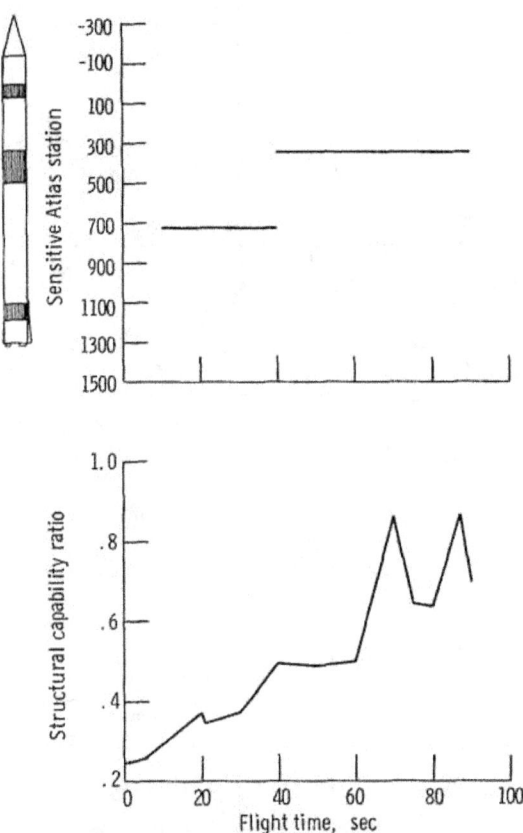

Figure VI-39. - Maximum predicted structural capability ratio, (total equivalent predicted bending moment)/(bending moment allowable), and sensitive station, AC-16.

94

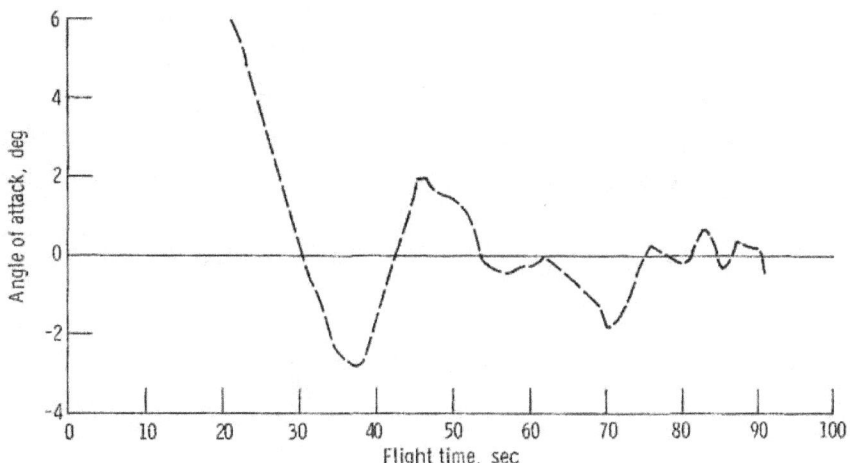

Figure VI-40. - Predicted pitch angles of attack, AC-16.

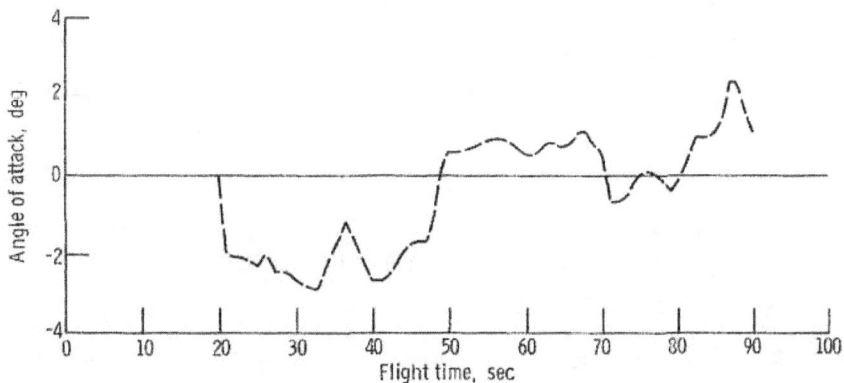

Figure VI-41. - Predicted yaw angles of attack, AC-16.

SEPARATION SYSTEMS

by Thomas L. Seeholzer, Charles W. Eastwood, and William M. Prati

Stage Separation

System description. - The Atlas-Centaur vehicle requires systems for Atlas booster engine section separation, Atlas-Centaur separation, and Centaur-spacecraft separation.

The Atlas booster engine stage separation system consists of 10 helium-gas-operated latch mechanisms. These latches (figs. VI-42 and VI-43) are located circumferentially around the Atlas aft bulkhead thrust ring at station 1133. An explosive valve supplies 2068-newton-per-square-centimeter (3000-psig) helium through the distribution manifold. When actuated, this results in the disengagement of the booster engine from the Atlas vehicle. Two tracks which extend from the thrust ring (fig. VI-44) are used to guide the booster engine section as it separates from the Atlas.

Atlas-Centaur staging systems (fig. VI-45) consist of a flexible, linear, shaped charge mounted circumferentially which severs the forward end of the interstage adapter at station 413; separation force is provided by eight retrorockets mounted on the aft end of the Atlas.

The OAO-II spacecraft is separated from Centaur by release of a four-segment band that clamps the spacecraft to the payload adapter. Release of the segmented band is accomplished by activation of four pyrotechnically operated latches. Separation force is provided by four mechanical spring assemblies, each having an 11.9-centimeter (4.7-in.) stroke, mounted on the payload adapter (fig. VI-46).

System performance. - Atlas booster engine section staging occurred 3 seconds after booster engine cutoff. This event was verified by data from instrumentation on the B-1 pitch actuator and from the vehicle axial (fine) accelerometer.

Atlas-Centaur staging was initiated at T + 236.4 seconds by the firing of the shaped charge which severed the interstage adapter at station 413. The eight retrorockets, mounted around the aft end of the Atlas, fired 0.1 second later to decelerate the Atlas and provide separation from the Centaur. Accelerometer and other data indicated that all eight retrorockets functioned as expected. Figure VI-47 shows the separation distance as a function of time after shaped-charge firing between the Atlas and the Centaur vehicle.

The yaw displacement gyros indicated that the Atlas rotated 0.39° about the yaw axis at the time the forward end of the interstage adapter cleared the Centaur engine nozzles. This resulted in a clearance loss in the minus yaw direction of 9.9 centimeter (3.9 in.) (fig. VI-47). The pitch rate gyros indicated 0.09° rotation about the pitch axis at the time the interstage adapter cleared the Centaur engine nozzles. This resulted in a clearance loss in the minus pitch direction of 2.3 centimeters (0.9 in.) (fig. VI-47). The resultant pitch and yaw clearance losses decreased the clearance (shown pictorially in fig VI-45) between the interstage adapter and Centaur engine nozzles from 27.9 to 24.1 centimeters (11 to 9.5 in.).

The latch pyrotechnics on the spacecraft clamping band were fired at T + 748.3 seconds, and the spacecraft separated from the Centaur vehicle.

Jettisonable Structures

System description. - The Atlas-Centaur vehicle jettisonable structures consists of hydrogen tank insulation panels, a nose fairing, and related separation systems.

The hydrogen tank insulation is made up of four polyurethane-foam-filled fiber glass honeycomb panels bolted together along the longitudinal axis to form a cylindrical cover around the Centaur tank. The panels are bolted at their aft end to a support on the Centaur vehicle. At the forward end, a circumferential Tedlar and fiber glass laminated cloth forms a seal between the panels and the base of the nose fairing at station 219. Separation of the four insulation panels is accomplished by firing the flexible, linear, shaped charges located at the forward, aft, and longitudinal seams. Immediately following shaped-charge firing, the panels rotate at their aft end about hinge points (fig. VI-48) because of the preload hoop tension, the center-of-gravity offset, and the in-flight residual purge pressure. The panels jettison free of the Centaur vehicle after approximately 45° of panel rotation on the hinge pins.

The vehicle nose fairing was a 3.05-meter- (10-ft-) diameter assembly approximately 12.2 meters (40 ft) long, consisting of a nonjettisonable section and a jettisonable section. The nonjettisonable section was composed of a fiber glass cylindrical barrel subassembly 1.83 meters (6 ft) long mounted on the forward flange of the Centaur tank and a metallic cylindrical fixed fairing subassembly 68.6 centimeters (27 in.) long bolted to the forward end of the barrel section. The jettisonable section consists of a metallic cylindrical split fairing subassembly 91.4 centimeters (36 in.) long bolted to a fiber glass cylindrical-to-conical configuration nose cone subassembly 8.84 meters (29 ft) long. This jettisonable section was assembled from two longitudinal halves joined by latch mechanisms. Each longitudinal half of the jettisonable section was attached to the fixed fairing subassembly with two hinges (fig. VI-49). The fairing hinges were each instrumented with biaxial strain gages (fig. VI-50) for jettison load measurements. Sheet cork on the exterior of the fiber glass section of the jettisonable nose fairing provided environmental protection to the spacecraft compartment and limited the fairing temperature to assure structural integrity. Separation of the nose fairing halves was accomplished by the firing of 10 pyrotechnically operated release latches along the split line and six release latches at the aft circumferential connection. Forces necessary to accomplish fairing jettison were applied by two preloaded compression spring assemblies mounted in the forward end, one in each fairing half. Upon release of the separation latches, the springs imparted the initial force to cause the fairing halves to pivot on their hinges. Each fairing half was free to leave the hinges after 30° of rotation (fig. VI-51).

System performance. - Flight data indicated that the four insulation panels were jettisoned satisfactorily. Insulation panel jettison sequence was commanded at T + 196.803 seconds. Data from axial accelerometers located on the spacecraft adapter indicated that the flexible, linear, shaped charges fired at T + 196.830 seconds. Event time data were provided by breakwire transducers to indicate 35° rotation of the panels. These breakwire transducers were attached to one hinge arm of each panel (fig. VI-52). Since these data were monitored on commutated channels, the panel 35° positive event times in the following table are mean times:

Panel location, quadrant	Instrumented hinge arm, quadrant	Event mean time, sec
I-II	I	T + 197.288
II-III	III	T + 197.394
III-IV	III	T + 197.296
IV-I	I	T + 197.348

Average angular velocities of the panels, assuming first motion at shaped-charge firing, were determined from the mean times of the 35° position. Panel velocities are compared in the following table with values from three previous flights:

[Minimum allowable rotation velocity, 40 deg/sec.]

Panel location, quadrant	Average rotation velocities from shaped-charge firing to mean time of 35° position, deg/sec			
	AC-16	AC-13	AC-14	AC-15
I-II	76.5	83.8	82.6	87.5
II-III	62.1	83.8	76.5	77.5
III-IV	75.2	82.6	84.3	82.5
IV-I	67.6	79.4	78.0	76.5

The rate for each panel was 10 to 20 percent lower than that observed on earlier flights. This was due in part to the lower jettison force resulting from a lower axial acceleration of the vehicle at panel jettison. A slightly lower fuel tank pressure at the event time resulted in a lower panel hoop tension and thereby also contributed to a lower jettison force. However, this combined force reduction did not account for the full difference in rotational rates. Another possible cause, and probably the major contributor, was variation within the allowed limits for panel installation pretension. The observed velocities were, however, considered to be well within design limits. Vehicle rates and dynamics at the event time indicated a completely satisfactory panel jettison sequence.

The nose fairing was jettisoned satisfactorily following issuance of the jettison command at T + 257.6 seconds. The pyrotechnically actuated unlatching mechanisms fired at T + 257.983 seconds and permitted the jettison springs to start rotation of the fairing halves. The times that fairing halves rotated $13.5°$, $15°$, and $45°$ are shown in the following table:

Fairing half	Angle of fairing rotation, deg		
	13.5	15	45
	Nose fairing rotation event time, sec		
+Y (without cap)	T + 258.494	T + 258.656	T + 259.628
-Y (with cap)	T + 258.494	T + 258.622	T + 259.627

The elapsed times of the fairing rotation events as determined from the flight data compared very well with data from the nose fairing jettison ground tests. These tests were full-scale, jettison system functional performance tests conducted at simulated altitude; the 1-g earth gravitational field approximated the vehicle longitudinal acceleration of 0.8 g expected at the time of nose fairing jettison. Comparable elapsed times of fairing rotation positions during flight and during ground test are summarized in the following table:

Fairing jettison event	Elapsed time from latch pyro-technic firing to nose fairing rotation position, sec			
	-Y fairing half (with cap)		+Y fairing half (without cap)	
	AC-16	Test	AC-16	Test
Unlatch firing	0	0	0	0
13.5° Rotation	.511	.585	.511	.579
15° Rotation	.639	.723	.673	.691
45° Rotation	1.644	1.667	1.645	1.660
Last detected hinge load	2.240	2.180	2.135	2.130

The strain gage flight data indicated that the hinge loads were as predicted and had close similarity to the fairing jettison test data.

The +Y fairing (without cap) commenced a buildup of hinge loads 0.025 seconds after fairing latch pyrotechnic firing. Hinge 1 reached a maximum axial compressive loading of 4670 newtons (1050 lbf) at latch firing plus 0.072 second. Hinge 2 experienced 6894 newtons (1550 lbf) of axial compressive load at latch firing plus 0.128 second. The loading of hinges 1 and 2 then went out of phase, indicating rocking of the fairing half on the hinges. This rocking motion was at approximately 2 hertz and continued with decreasing magnitude until separation of the hinge at latch pyrotechnic firing plus 2.13 seconds. The out-of-phase loading on the hinges is shown in figure VI-53. Oscillations at a frequency of 25 hertz were evident on both hinges for the first 1.5 seconds of rotation. Neither hinge experienced any axial tension loads, but the loading on both hinges decreased to zero several times as a result of the rocking action. Radial loads on the +Y fairing hinges indicated the rocking effect also. Maximum radial loads on both hinges were always less than 2224 newtons (500 lbf).

On the -Y fairing (with cap), hinge 4 began to experience an axial compressive load at latch firing plus 0.025 second. Oscillations at a frequency of 20 hertz were evident for 0.5 second of rotation. A maximum axial load of 9341 newtons (2100 lbf) occurred at latch pyrotechnic firing plus 0.350 second.

Hinge 3 had essentially no axial load buildup for 0.55 second of fairing rotation. The data indicated that the -Y fairing also rocked on the hinges at a frequency of 2 hertz from first rotation to hinge separation at latch pyrotechnic firing plus 2.24 seconds. Oscillations at a frequency of 25 hertz also were evident on this fairing. The maximum axial compressive load on hinge 3 was 4893 newtons (1100 lbf) at 0.605 second after latch pyrotechnic firing. This hinge did experience some small axial tension loads which did not exceed 445 newtons (100 lbf). Hinge 4 axial loads did diminish to zero several times but never went into tension from the rocking action. Radial loads on this fairing half also showed the rocking effect. On hinges 3 and 4 the radial loads were not over 2669 newtons (600 lbf) at any time.

Jettison of the fairing did result in some vehicle transients, principally roll oscillations. The rocking action of the fairing halves was in phase on opposite hinges and produced roll torques. This resulted in roll oscillations of approximately the same frequency and duration as the hinge load fluctuations, 2 hertz for 2 seconds.

The hinge loads at nose fairing jettison during flight were well within the design limits and were similar to those generated in the ground jettison test as shown in the following table:

Type of load	Units	Maximum hinge loads at nose fairing jettison, AC-16							
		During flight[a]				During ground test[b]			
		Hinge							
		1	2	3	4	1	2	3	4
Maximum axial compression	N	4670	6894	4893	9341	6405	8006	5338	6316
	lbf	1050	1550	1100	2100	1440	1800	1200	1420
Maximum radial acting inboard	N	1801	1312	2513	1023	1334	2268	1557	1913
	lbf	405	295	565	230	300	510	350	430
Maximum radial acting outboard	N	1557	2068	1913	2291	2046	2180	1690	1824
	lbf	350	465	430	515	460	490	380	410
		Combined hinges, fairing half -							
		+Y		-Y		+Y		-Y	
Maximum axial compression	N	10 898		9118		12 677		11 565	
	lbf	2 450		2050		2 850		2 600	
Maximum radial acting inboard	N	1 957		3269		1 423		3 247	
	lbf	440		735		320		730	
Maximum radial acting outboard	N	2 669		2291		3 158		3 514	
	lbf	600		515		710		790	

[a]Only hinge 3 experienced axial tension loads; maximum value was 445 N (100 lbf).

[b]These values are the maximum values obtained during the series of three ground tests.

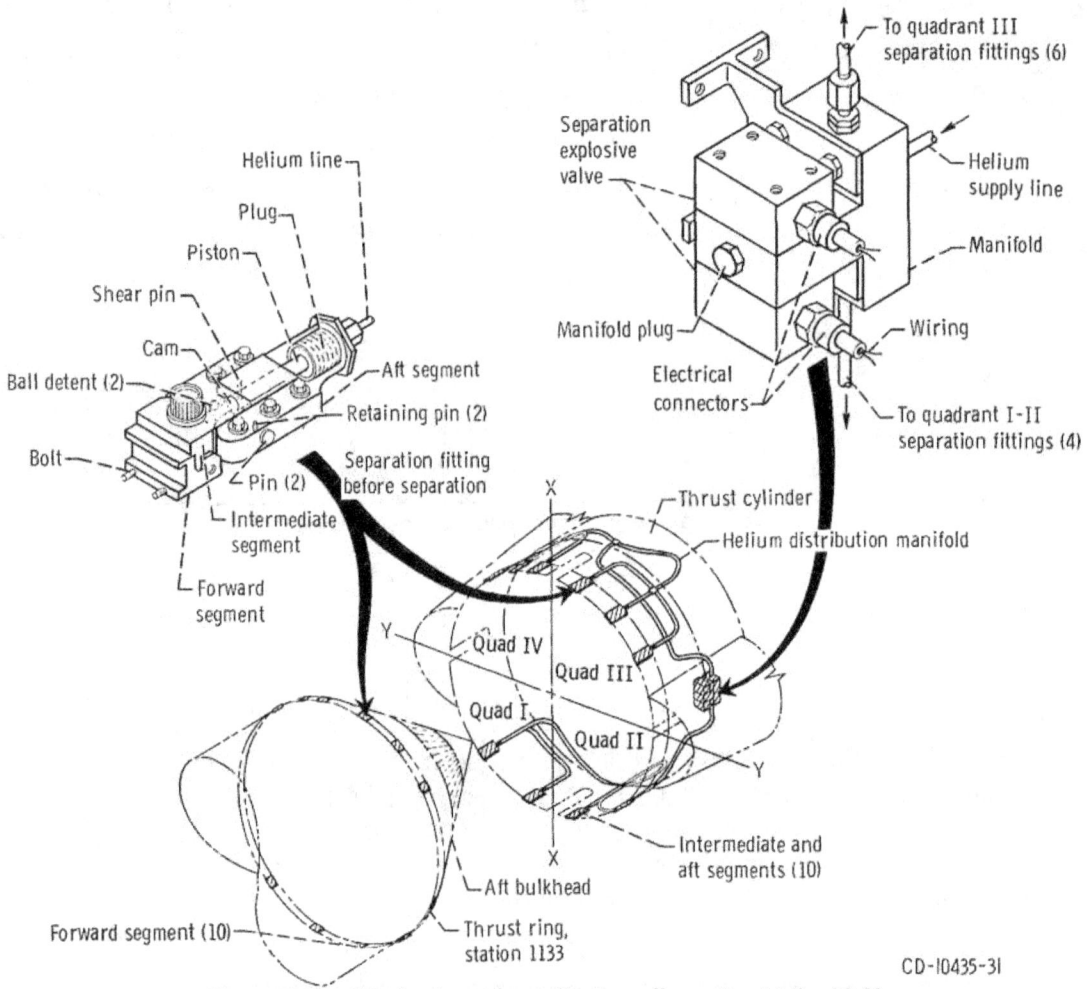

Figure VI-42. - Atlas booster engine section separation system details, AC-16.

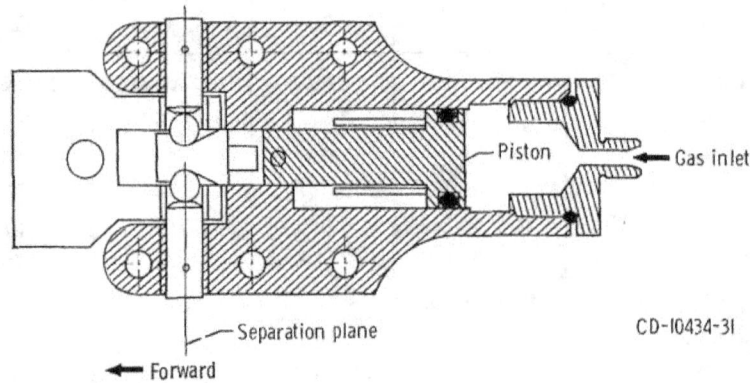

Figure VI-43. - Atlas booster engine section separation fitting, AC-16.

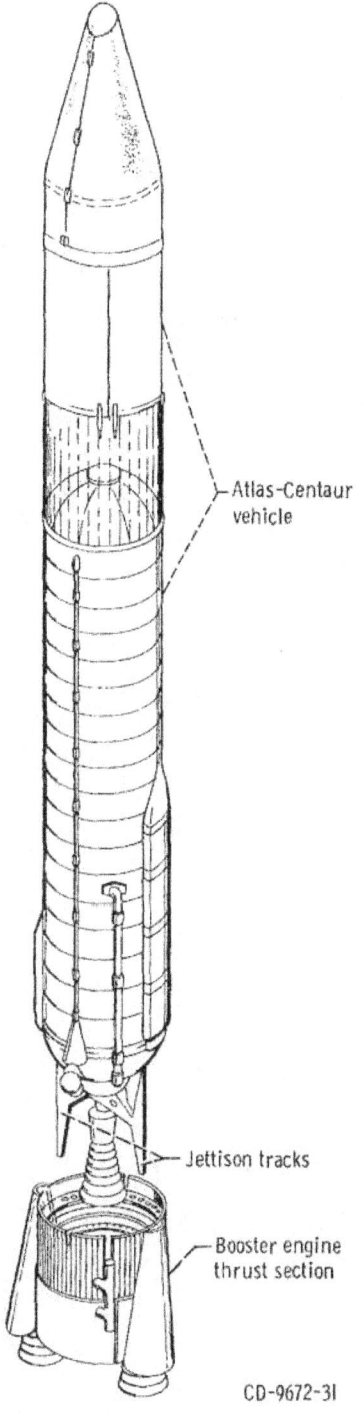

Atlas-Centaur vehicle

Jettison tracks

Booster engine thrust section

CD-9672-31

Figure VI-44. – Atlas booster engine section staging system, AC-16.

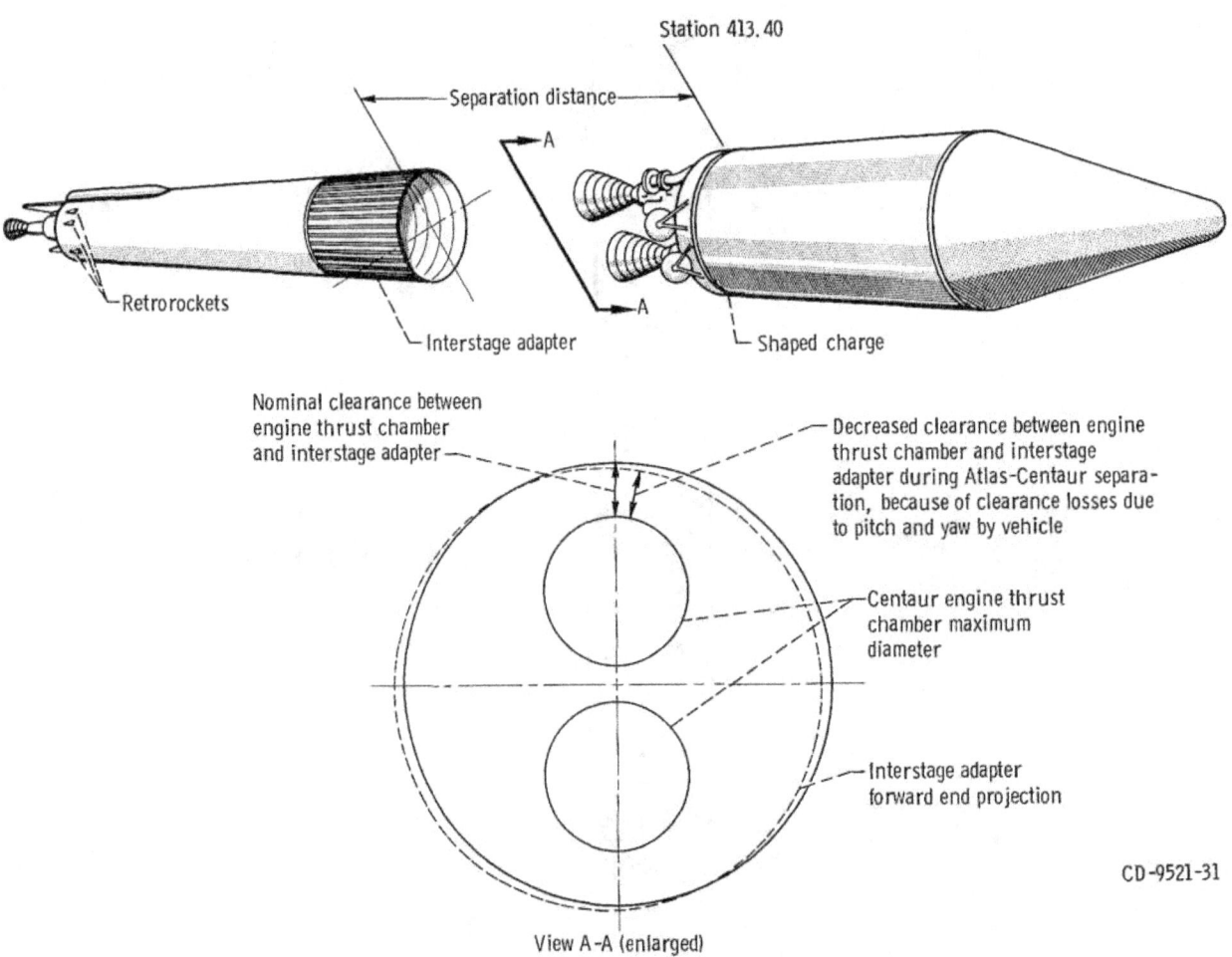

Station 413.40

Separation distance

A

A

Retrorockets

Interstage adapter

Shaped charge

Nominal clearance between engine thrust chamber and interstage adapter

Decreased clearance between engine thrust chamber and interstage adapter during Atlas-Centaur separation, because of clearance losses due to pitch and yaw by vehicle

Centaur engine thrust chamber maximum diameter

Interstage adapter forward end projection

CD-9521-31

View A-A (enlarged)

Figure VI-45. – Atlas-Centaur separation system, AC-16.

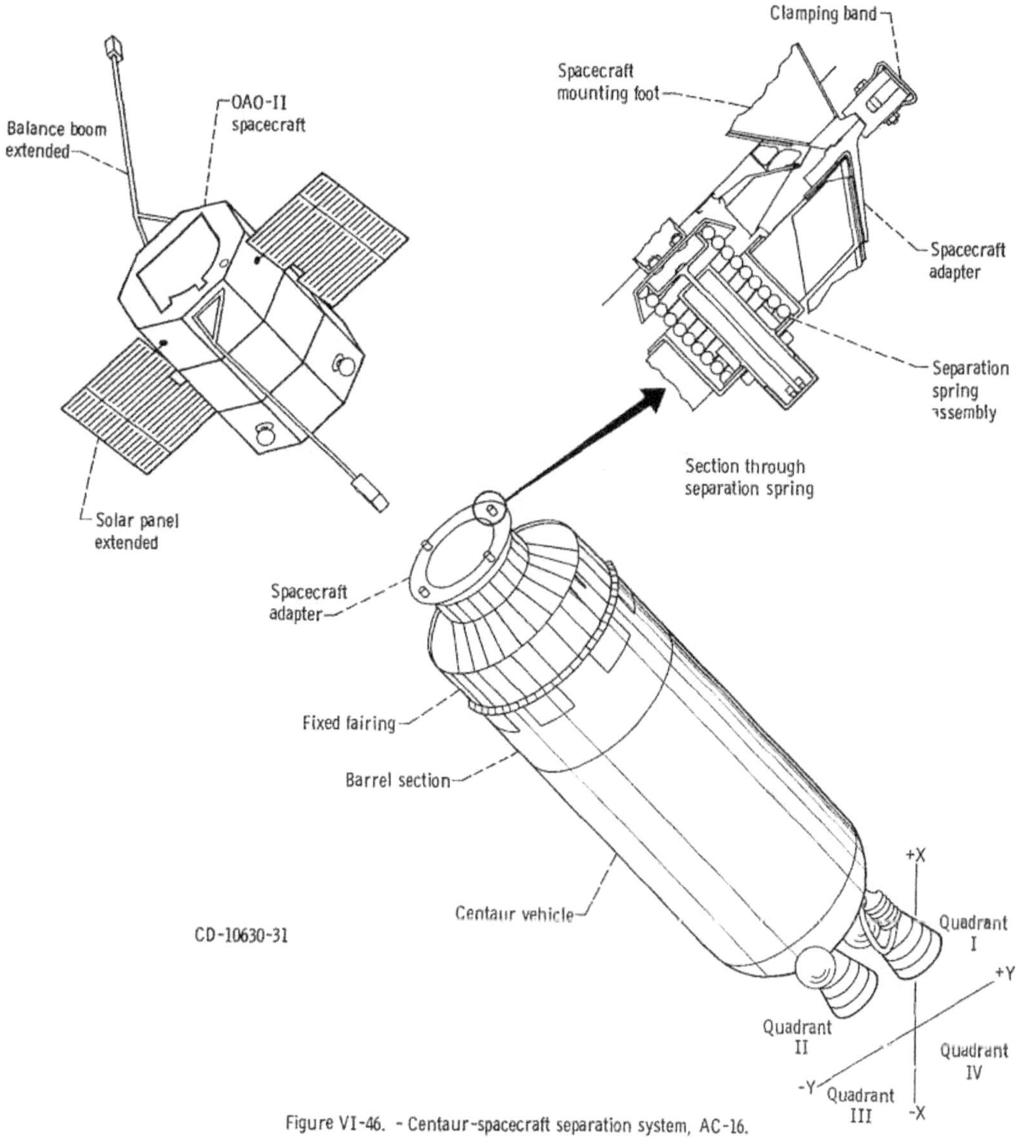

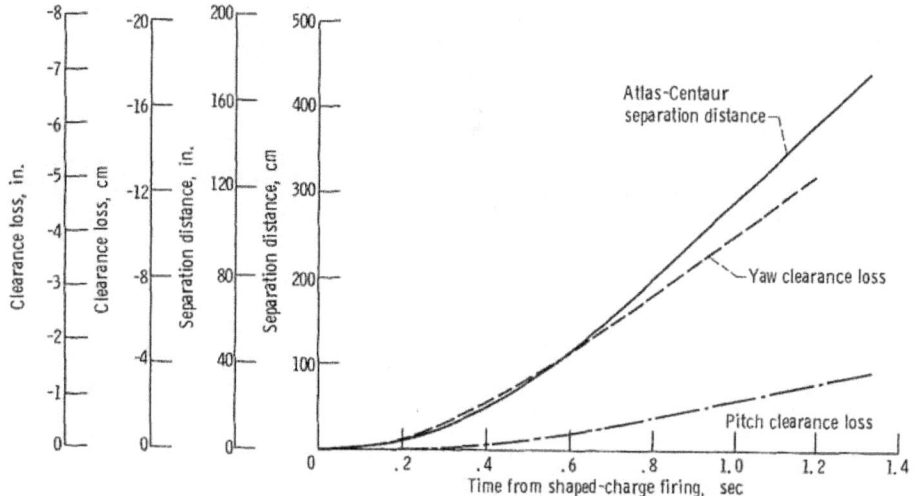

Figure VI-46. - Centaur-spacecraft separation system, AC-16.

Figure VI-47. - Atlas-Centaur separation distances, AC-16. All clearance losses referenced to forward end of interstage adapter (station 413).

105

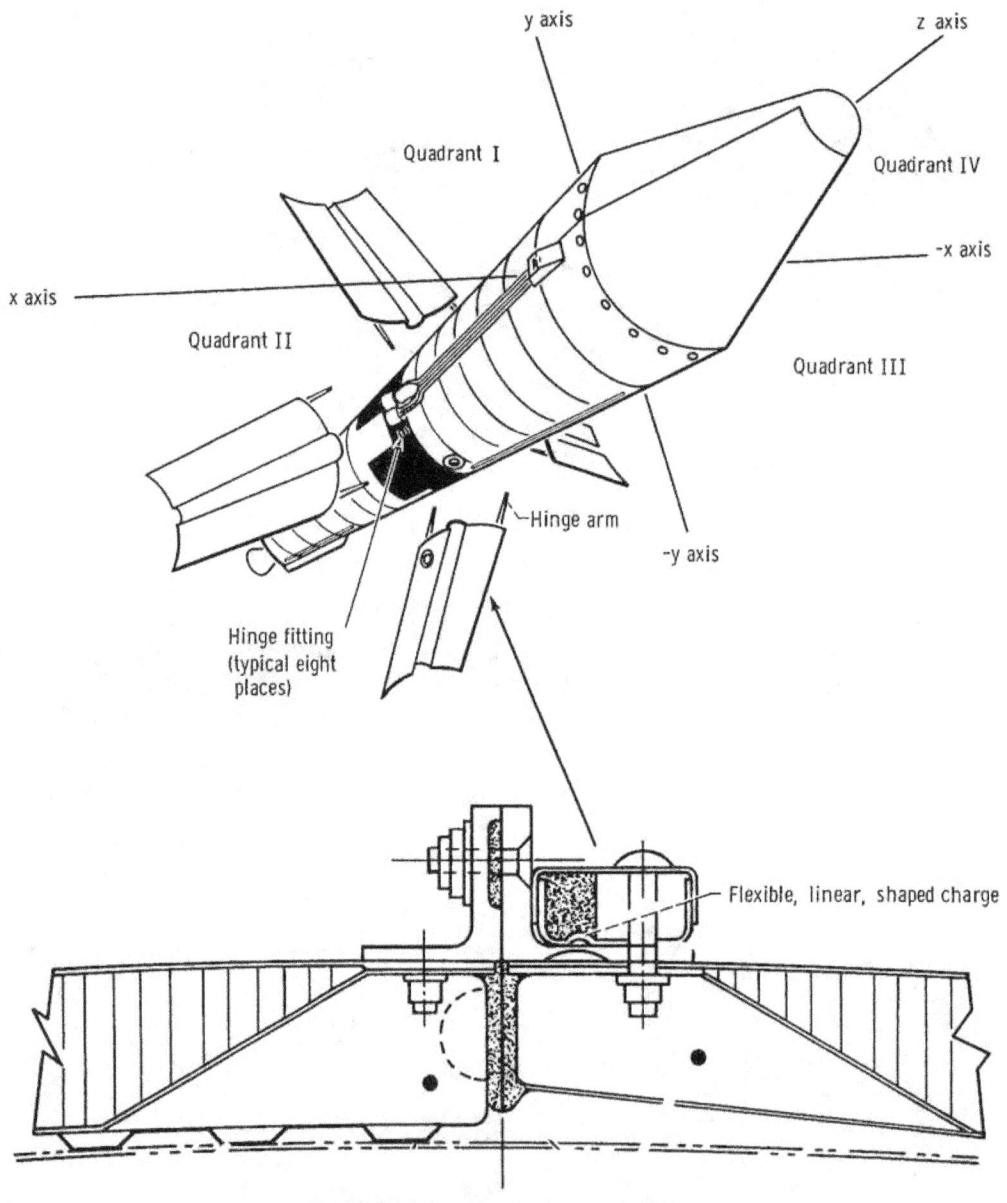

Longitudinal seam shaped-charge installation

Figure VI-48. - Hydrogen tank insulation jettison system, AC-16.

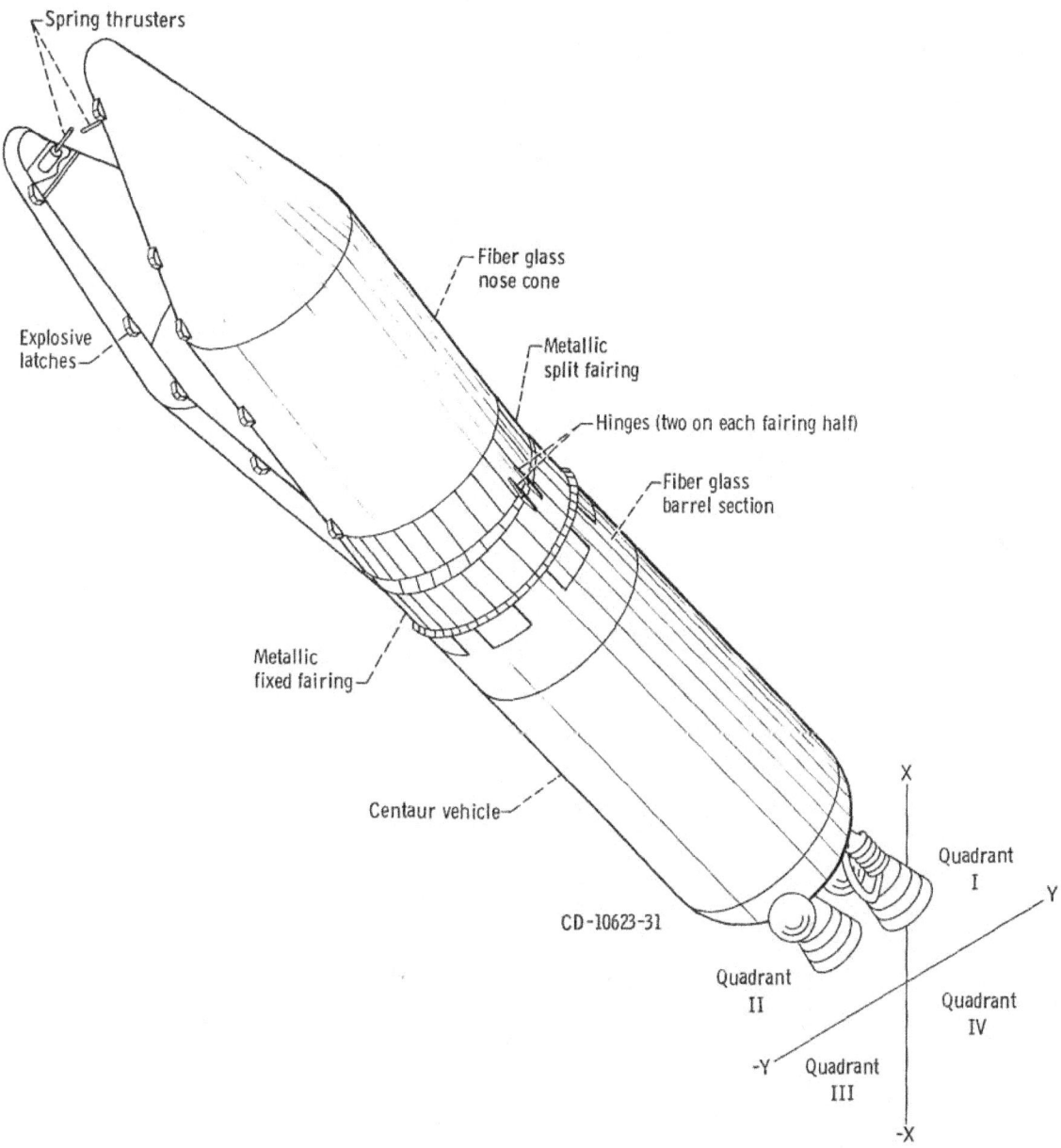

Spring thrusters

Fiber glass
nose cone

Metallic
split fairing

Hinges (two on each fairing half)

Explosive
latches

Fiber glass
barrel section

Metallic
fixed fairing

Centaur vehicle

CD-10623-31

X

Quadrant
I

Y

Quadrant
II

Quadrant
IV

-Y   Quadrant
III

-X

Figure VI-49. - Nose fairing jettison system, AC-16.

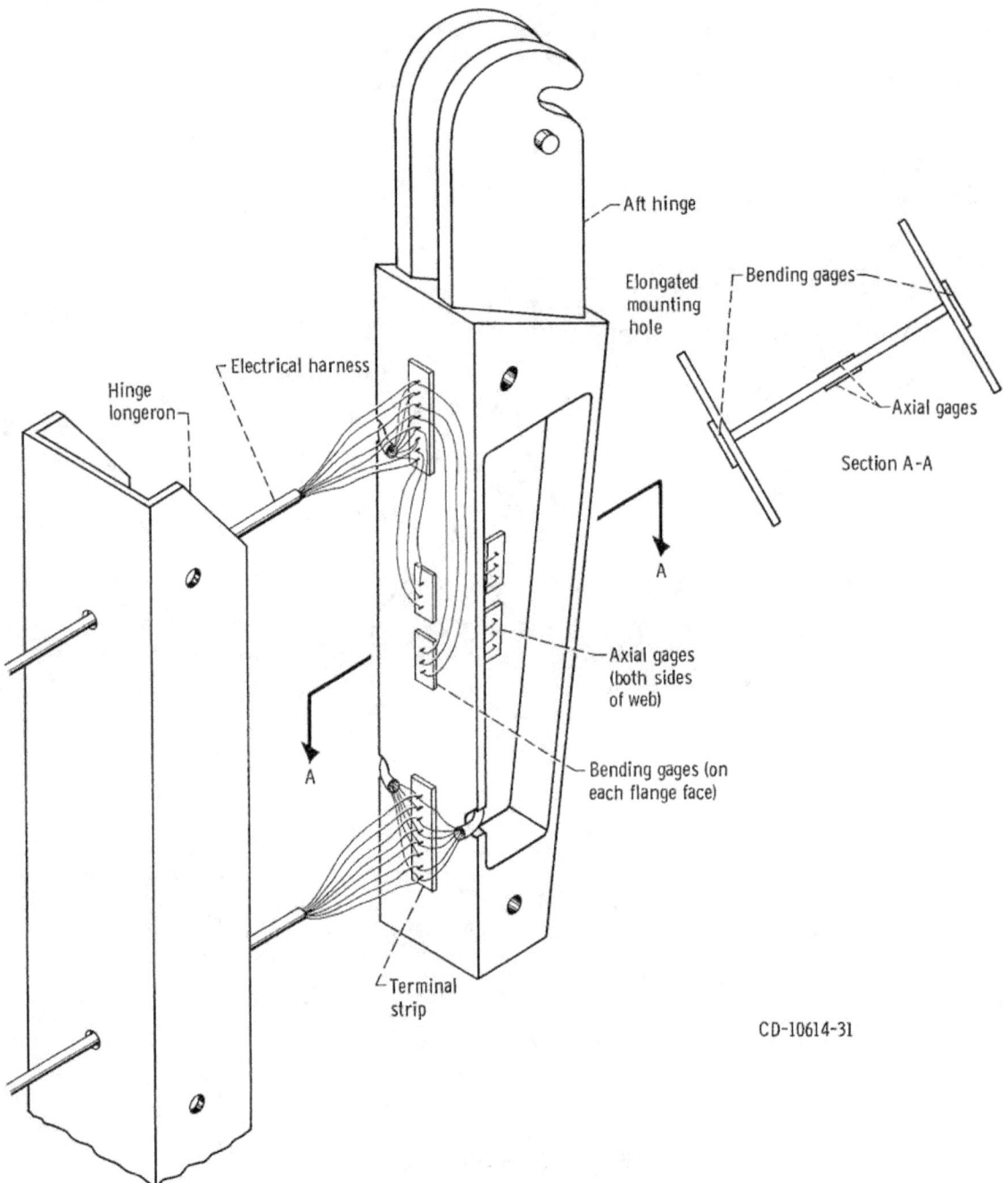

Figure VI-50. – Nose fairing hinge strain gages, AC-16. (Typical for four hinges.)

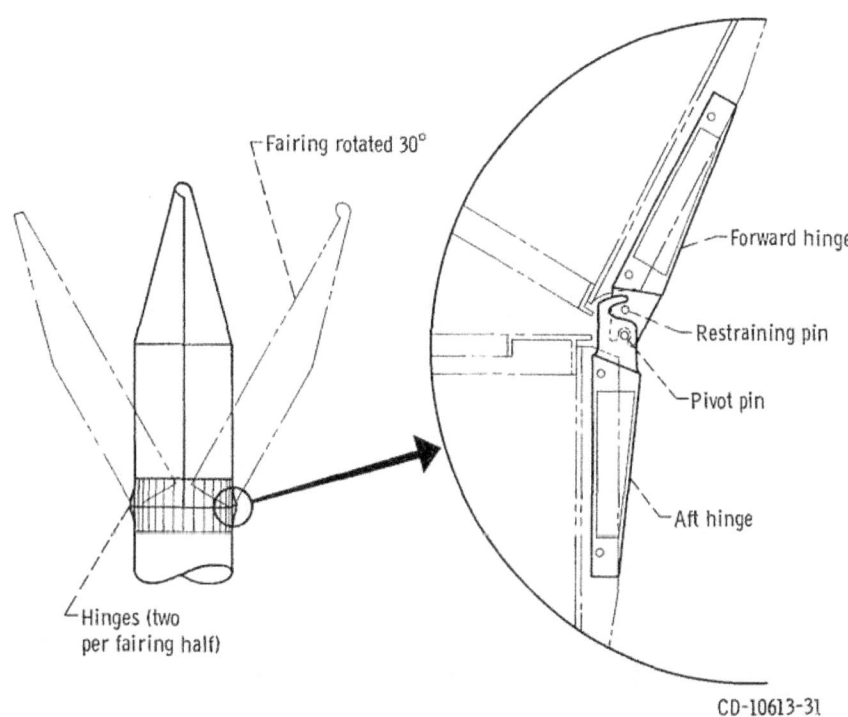

Figure VI-51. - Nose fairing hinges, AC-16.

CD-10613-31

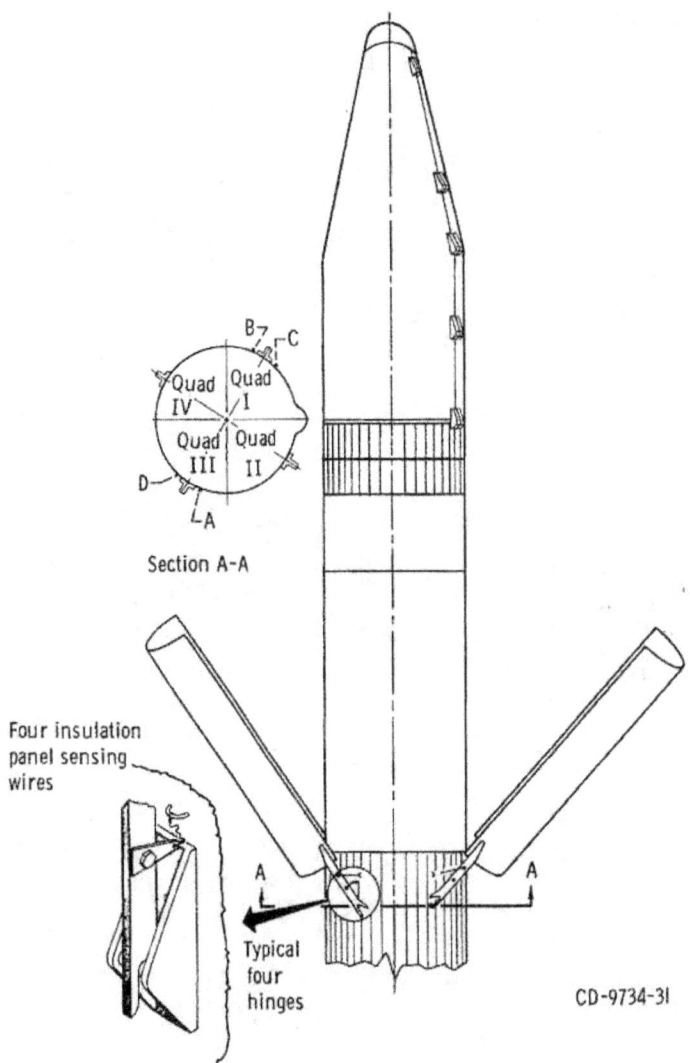

Section A-A

Four insulation
panel sensing
wires

Typical
four
hinges

CD-9734-3I

Figure VI-52. - Insulation panel breakwire locations, AC-16.

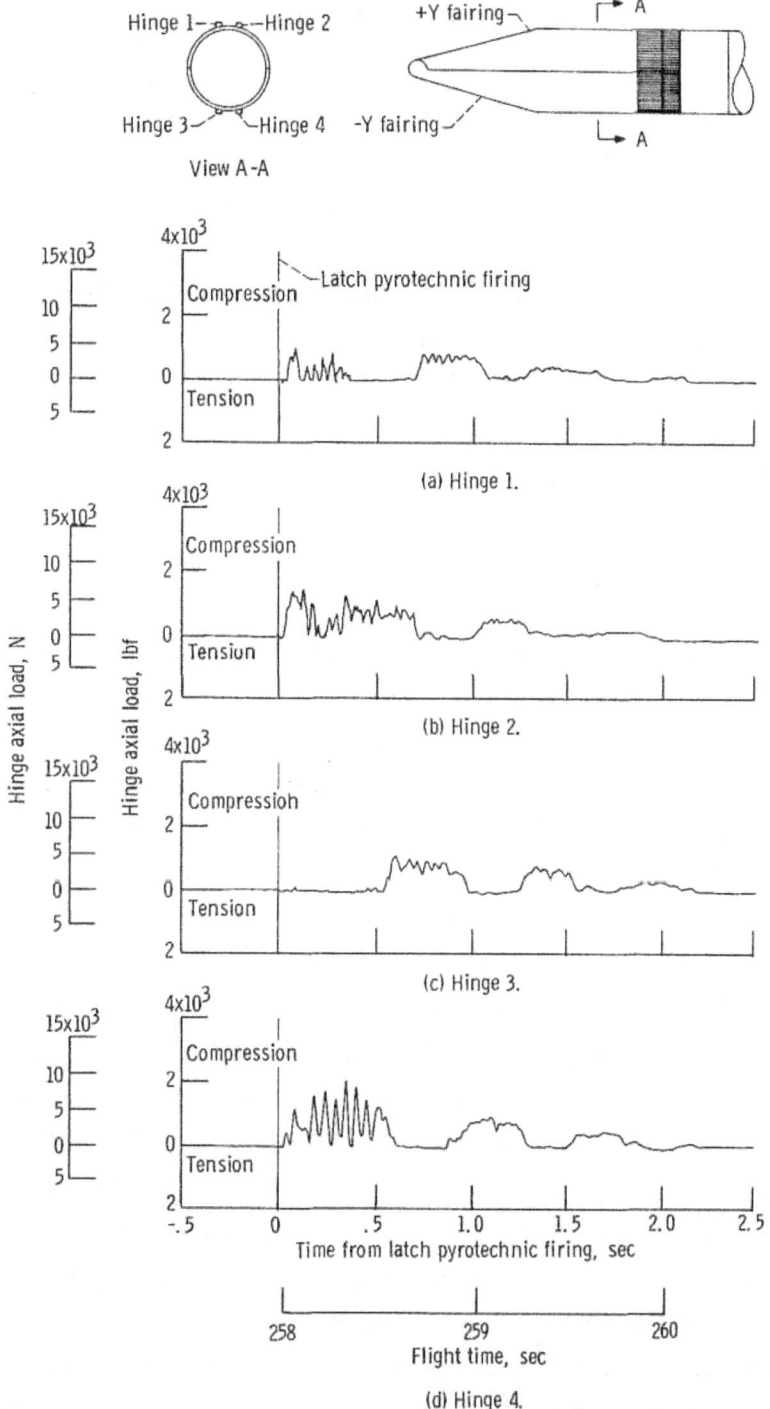

Figure VI-53. - Nose fairing hinge axial loads at jettison, AC-16.

111

ELECTRICAL SYSTEMS

by John M. Bulloch and John B. Nechvatal

Power Sources and Distribution

System description - Atlas. - The power supply consists of a power changeover switch, one main battery, one telemetry battery, two independent range safety command (vehicle destruct) system batteries, and a three-phase 400-hertz rotary inverter.

System performance - Atlas. - Transfer of the Atlas electrical load from external to internal power was accomplished by the main power changeover switch at T - 2 minutes. Performance of the Atlas electrical system was normal throughout the flight. Voltages and current levels furnished to the dependent systems were within specification limits.

The Atlas main battery supplied the requirements of the user systems at normal levels (25 to 30 V). The battery voltage was 28.1 volts at lift-off and 28.4 volts at sustainer engine cutoff. This is a normal rise of voltage, and is caused by battery temperature increase.

The telemetry battery and the two range safety command batteries provided normal voltage levels throughout Atlas flight. The voltages at lift-off were 28.4 volts for the telemetry system, and 29.4 and 29.7 volts for the two range safety command systems.

The Atlas rotary inverter supplied 400-hertz power within specified voltages and frequency limits. The voltage at lift-off was 115.1 volts and decreased to 114.7 volts at the end of Atlas powered flight (234.5 sec). The inverter frequency at lift-off was 401.8 hertz and increased to 402.8 hertz by the end of Atlas flight. The gradual rise in frequency is typical of the Atlas inverter. The required frequency difference range of 1.3 to 3.7 hertz between the Atlas and Centaur inverter frequencies was properly maintained. Operation in this range prohibits the possibility of generating undesirable beat frequencies within the flight control system, thereby precluding the chance of resonant excitation of propellant "slosh" modes or natural frequencies of the vehicle structure.

System description - Centaur. - The electric power system consists of a power changeover switch, a main battery, two independent range safety command (vehicle destruct) batteries, two pyrotechnic system batteries, and a solid-state inverter supplying 400-hertz current to the guidance, flight control, and propellant utilization systems.

System performance - Centaur. - Performance of the Centaur electrical system was satisfactory throughout the flight. Transfer of the Centaur electric load from external power to the internal battery was accomplished at T - 4 minutes by the power changeover switch in 250 milliseconds (specification value, 2.00 sec max.). The maximum voltage excursion at power transfer was 0.5 volt, which is considered negligible.

The main battery voltage at lift-off was 27.9 volts. It decreased to 27.2 volts at Centaur main engine start (T + 246.0 sec), and rose to a steady-state level of 28 volts during Centaur powered flight. Following main engine cutoff the voltage increased to 28.2 volts and remained constant throughout the coast phase. These values are consistent with normal battery regulation. The main battery voltage was 27.5 volts, with a sustained vehicle load of 50 amperes at the loss of telemetry data (T + 8100 sec). An approximate total of 115 ampere-hours had been expended from the battery. (Rated battery capacity is 100 A-hr.)

The flight current profile (fig. VI-54) was normal and correlated closely with tests prior to launch. At lift-off the current was 44 amperes, increasing to 61 amperes at Centaur engine start (T + 246.0 sec).

A series of current disturbances of 0.5 ampere peak-to-peak was observed, starting at T + 682.4 seconds. These small disturbances were not detrimental; their cause has not been determined. Simultaneous disturbances were also observed on the six spacecraft accelerometers at precisely the same time.

The voltages of the nose fairing pyrotechnic batteries were 34.85 and 34.9 volts at lift-off. These batteries had a higher capacity than those used in previous Centaur flights in order to meet the higher power requirements of the AC-16 nose fairing pyrotechnic system. Proper operation of the pyrotechnic system batteries and associated relays was verified by instrumentation and by the successful jettison of the insulation panels and the nose fairing.

Performance of the two range safety command system batteries was satisfactory. At lift-off the two battery voltages were 32.6 and 33.6 volts. The minimum specification limit is 30 volts.

The Centaur inverter operated satisfactorily throughout the flight. Telemetered voltage levels compared closely to values recorded during preflight testing. The inverter phase voltages at lift-off were as follows: Phase A, 115.3 volts; Phase B, 115.5 volts; and Phase C, 115.4 volts. Voltage changes during flight were small and well within expected values. Inverter frequency remained constant at 400.0 hertz throughout the flight. Inverter skin temperature was 304.2 K (88° F) at lift-off and reached a high of 360 K (188° F) by T + 8100 seconds.

Instrumentation and Telemetry

System description - Atlas. - The Atlas telemetry system (fig. VI-55) consists of a radiofrequency telemetry package, two antennas, a telemetry battery, and transducers.

It is a Pulse Amplitude Modulation/Frequency Modulation/Frequency Modulation (PAM/FM/FM) telemetry system and operates at a carrier frequency of 229.0 megahertz. The PAM technique used on all Atlas-Centaur commutated (sampled) channels makes possible a larger number of measurements on one subcarrier channel. This increases the data handling capability of the telemetry system. The FM/FM technique uses analog values from transducers to frequency modulate the subcarrier oscillators, which, in turn, frequency modulate the main carrier (radiofrequency link).

System performance - Atlas. - The 107 operational measurements shown in table VI-IX were transmitted through two antennas located on the two equipment pods. Satisfactory telemetry coverage was obtained to T + 813 seconds (fig. VI-56). All measurements provided useful data. Minor anomalies were experienced with the following three measurements:

(1) Sustainer hydraulic return line pressure measurement (Alt601P): Erratic data variations and "spiking" were experienced from approximately T + 37 to T + 57 seconds and from T + 92 to T + 158 seconds. The data at all other times were at expected levels and trends. The cause of the data variations is not known.

(2) Sustainer engine yaw position measurement (As256D): This measurement exhibited a bias condition of approximately 10-percent Information Band Width (IBW) below the expected 50 percent level prior to lift-off and during the flight. Other than the bias shift, data levels and trends were satisfactory. The cause of the bias shift was probably slippage of the sustainer engine yaw position transducer adjustment.

(3) Insulation panel 35°-of-rotation breakwire measurements (AA205X/AA208X) for quadrants III-IV and IV-I: The measurement AA205X indicated 35° of rotation of the quadrant III-IV insulation panel at T + 197.31 seconds (61-percent IBW). The 61-percent level also indicates the quadrant IV-I insulation panel had not rotated 35°. At T + 197.328 seconds the measurement AA208X indicated 35° of rotation of the quadrant IV-I insulation panel (37-percent IBW). The 37-percent IBW also indicated the quadrant III-IV insulation panel had not rotated. Subsequent data samples were at the correct level (80-percent IBW). The 80-percent IBW indicates both the quadrant III-IV and the quadrant IV-I insulation panels had rotated 35°. It is theorized that the quadrant III-IV insulation panel breakwire transducer opened, then momentarily closed, then opened again.

System description - Centaur. - The Centaur telemetry system (fig. VI-57) consists of two radiofrequency telemetry packages, one antenna (fig. VI-58), and transducers. It is a PAM/FM/FM system which operates at frequencies of 225.7 and 259.7 megahertz for radiofrequency telemetry package 1 (RF1) and radiofrequency telemetry package 2 (RF2), respectively. The RF1 telemetry package was used for vehicle operational measurements, and the RF2 telemetry package was installed in order to define (1) the low-frequency rigid-body vibration environment at the spacecraft interface,

(2) axial and radial loads on the nose fairing hinges, (3) vehicle bending modes in the fixed fairing, (4) nose fairing jettison separation rates and clearances, and (5) spacecraft acoustic environment.

System performance - Centaur. - A total of 181 measurements (table VI-X) were transmitted to the ground stations. Anomalies were experienced with the following measurements:

(1) Centaur engine C-1 thrust chamber jacket temperature measurements (CP63T) went off scale high at T + 264 seconds. This indicates an open circuit in either the transducer or the harnessing. No data were retrieved beyond this time.

(2) Centaur engine C-2 fuel pump case temperature measurement (CP122T) exhibited a slow response to temperature changes. This malfunction has been observed on previous flights, and the problem is believed to be caused by the temperature transducer losing its bond to the fuel pump case.

(3) Centaur engine C-2 hydraulic power package pressure measurement (CH3P) exhibited negative "spiking" of up to 8-percent IBW during the powered portion of flight. This is indicative of wiper arm lift-off in the transducer. No data were lost because of this malfunction.

(4) The six accelerometers mounted on the payload adapter exhibited small disturbances that coincided with a series of 0.5-ampere "spikes" on the battery current monitor. The cause of this anomaly has not been determined.

(5) Guidance digital data after approximately 92 minutes into the flight showed intermittent computer telemetry bits occurring during the telemetry work marker and immediately following some data words. The problem has been isolated to the telemetry output circuitry in the guidance computer being temperature sensitive. No data were lost because of this malfunction.

Satisfactory telemetry coverage was obtained to T + 8256 seconds as shown in figure VI-59. Location of the receiving stations are shown in figure VI-60.

Noisy data were experienced at Canary Island from approximately T + 1079 to T + 1161 seconds. Tananarive had intermittent noise from approximately T + 2351 to T + 2461 seconds. Carnarvon did not record radiofrequency link 1, channels 1, 2, 3, and 4, because of the ground equipment problems.

Tracking System

System description. - The tracking system consisting of an airborne C-band radar beacon subsystem (fig. VI-60), with associated radar ground stations (fig. VI-61), provided real-time position and velocity data to the range safety tracking system. These data were also used by the Manned Space Flight Network (MSFN) for assistance in acquisition of the spacecraft and for guidance and flight trajectory data analysis. The airborne equipment includes a lightweight transponder, a circulator (to channelize receiving and sending signals), a power divider, and two antennas located on opposite sides of the Centaur vehicle. The locations of the C-band antennas are shown in figure VI-58.

System performance. - The C-band radar tracking was satisfactory; coverage was obtained up to T + 7998 seconds, as indicated in figure VI-62. Antigua, Canary Island, Tananarive, Carnarvon, and Hawaii provided data to verify orbital conditions. Merritt Island, Patrick, Grand Bahama Island, Grand Turk, and Bermuda provided coverage during Atlas-Centaur powered flight and data for orbit calculation. Merritt Island, Bermuda, and Canary Island also provided second orbital pass coverage.

Range Safety Command System

System description. - The Atlas and Centaur stages each contained independent vehicle destruct systems. Each system included redundant receivers, a power control unit, destructor, two antennas, and batteries which operate independently of the main vehicle power system. The location of the Centaur range safety antennas is shown in figure VI-58. Block diagrams of the Atlas and Centaur vehicle destruct systems are shown in figures VI-63 and VI-64, respectively. These systems were designed to function simultaneously upon command from the ground stations.

The Atlas and Centaur vehicle destruct systems provide a highly reliable means of shutting down the engine only, or shutting down the engines and destroying the vehicle, if it had left the safe flight corridor. To destroy the vehicle, the propellant tanks would be ruptured with an explosive charge and the liquid propellants of the Atlas and Centaur dispersed.

System performance. - The Atlas and Centaur vehicle destruct systems were prepared to execute destruct command throughout the flight. Engine cutoff or destruct commands were not sent by the range transmitter. The command from the Bermuda ground station to disable the range safety command system shortly after Centaur main engine cutoff was properly received and executed. Figure VI-65 depicts ground transmitter coverage to support the vehicle destruct systems.

The receiver signal strength measurements indicated a satisfactory received signal strength throughout the flight, with one exception at approximately T + 107 to T + 112 seconds. Just prior to the switching of the range transmitter from Cape Kennedy to Grand Bahama, the signal strength to all four receivers dropped below the 12-decibel gain margin above threshold required by the range for range safety operation. It is believed the signal strength decrease was due to inadequate signal strength at the vehicle, which was caused by the range not switching the Cape Kennedy range transmitter from low power (600 W) to high power (10 000 W). Upon switching to the Grand Bahama transmitter at T + 112 seconds, the signal strength at all four receivers increased substantially. Telemetered data indicated that both the Centaur receivers were deactivated at approximately T + 705 seconds, thus confirming that the disable command was transmitted from the Bermuda station.

TABLE VI-IX. - ATLAS MEASUREMENT SUMMARY, AC-16

System	Measurement type										Totals
	Vibration	Acceleration	Rotation rate	Displacement	Pressure	Frequency	Rate	Temperature	Voltage	Discretes	
Airframe	2	1			3			2		4	12
Range safety									3	1	4
Electrical						1			4		5
Pneumatics					7			1			8
Hydraulics					6						6
Dynamics		1								2	3
Propulsion			3	3	18			6		6	36
Flight control				11			3		4	11	29
Propellants		1			2				1		4
Totals	2	3	3	14	36	1	3	9	12	24	107

TABLE VI-X. - CENTAUR MEASUREMENT SUMMARY, AC-16

System	Measurement type														Totals
	Acoustical	Strain	Acceleration	Rotation rate	Current	Displacement	Vibration	Pressure	Frequency	Rate	Temperature	Digital	Voltage	Discretes	
Airframe		11	7			1					2			6	27
Range safety													2	3	5
Electrical					1				1		2		4		8
Pneumatics								6			4			2	12
Hydraulics								2			2				4
Guidance											1	1	16		18
Propulsion				4				12			28			10	54
Flight control										3			4	30	37
Propellant						2							2		4
Spacecraft	1						3	1			1			6	12
Totals	1	11	7	4	1	3	3	21	1	3	40	1	28	57	181

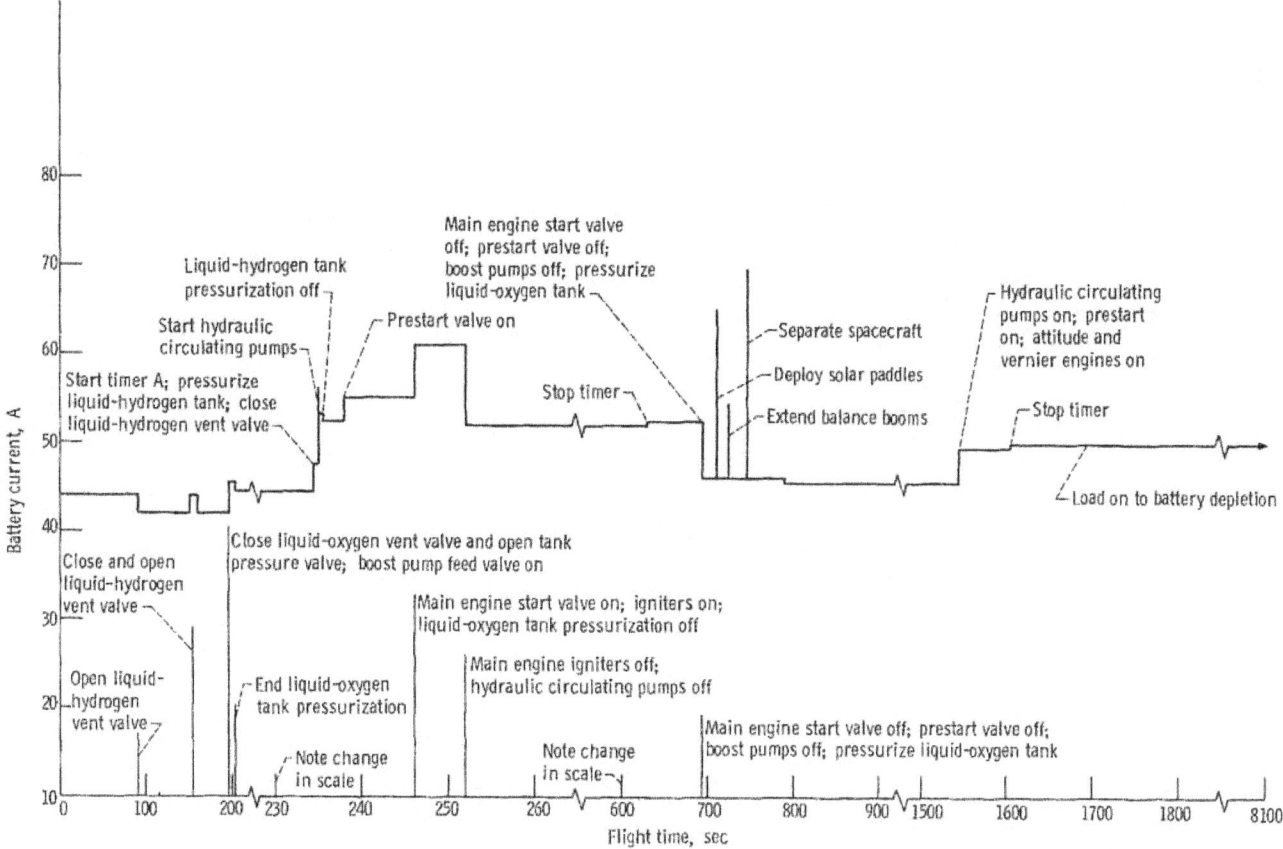

Figure VI-54. - Centaur main battery current profile, AC-16.

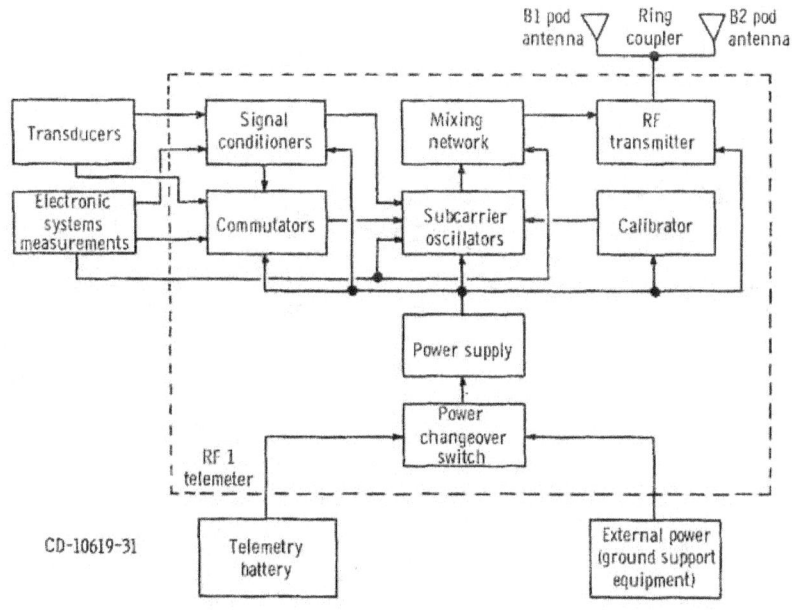

CD-10619-31

Figure VI-55. - Atlas telemetry system block diagram, AC-16.

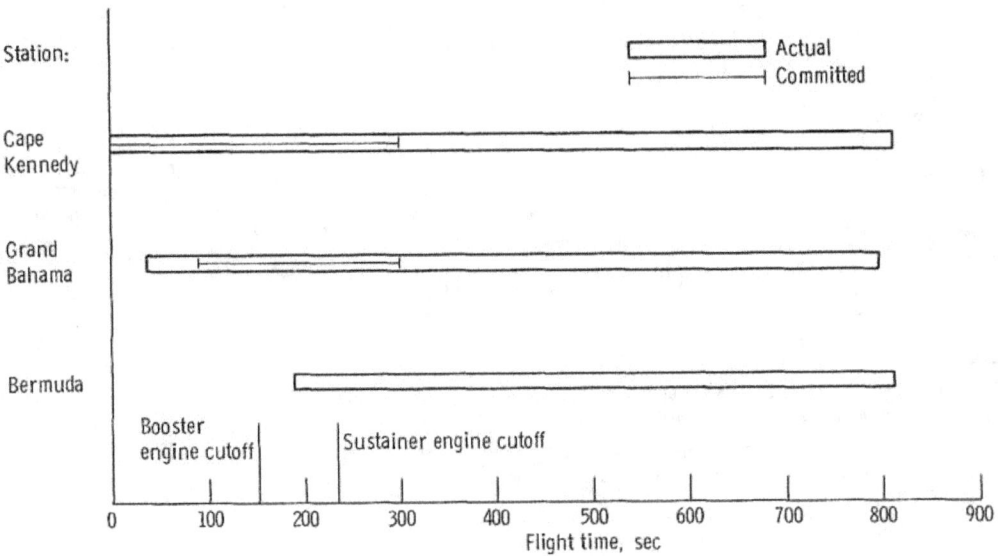

Figure VI-56. - Atlas telemetry coverage, AC-16.

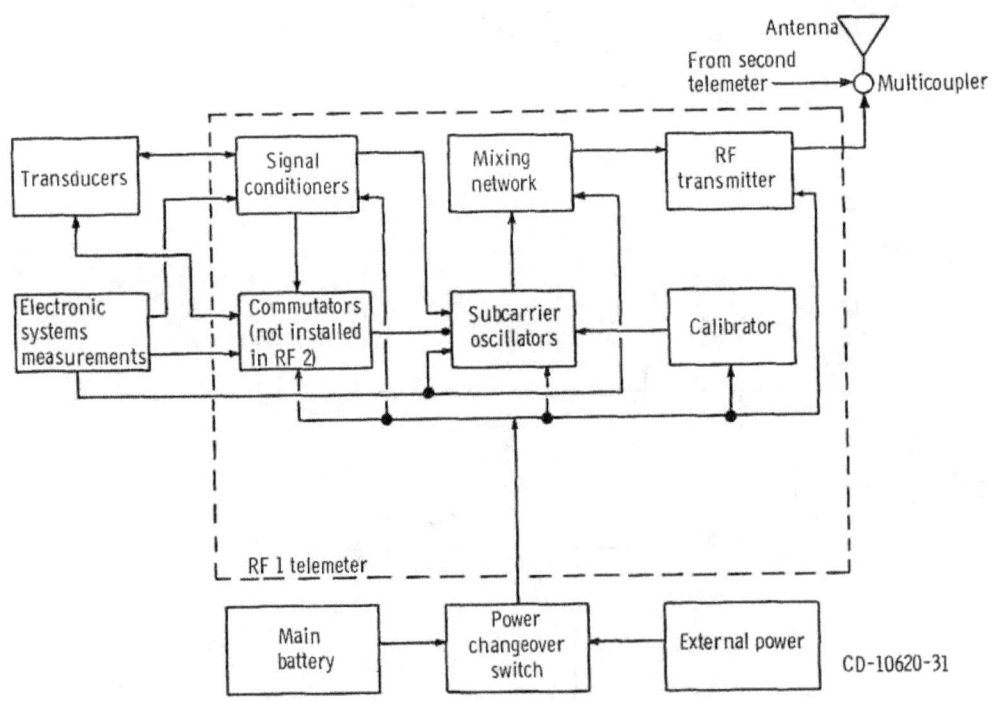

CD-10620-31

Figure VI-57. - Centaur telemetry system, typical both telemeters, AC-16.

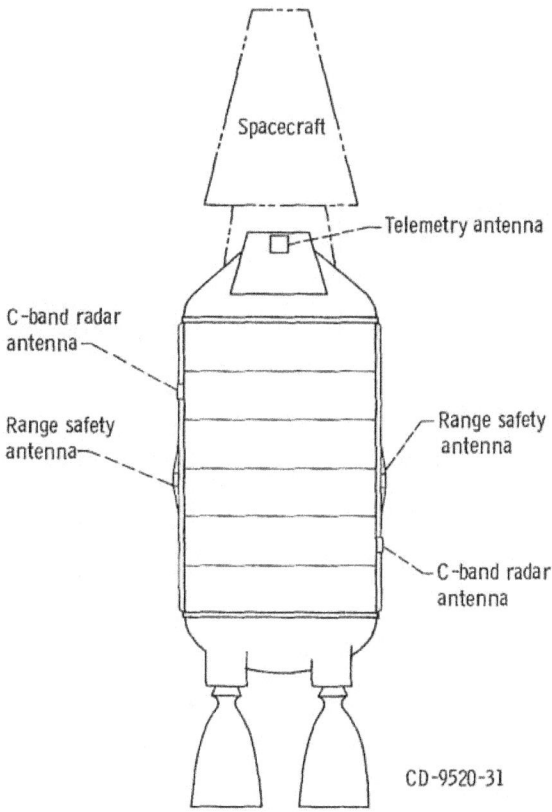

Figure VI-58. – Location of Centaur antennas, AC-16.

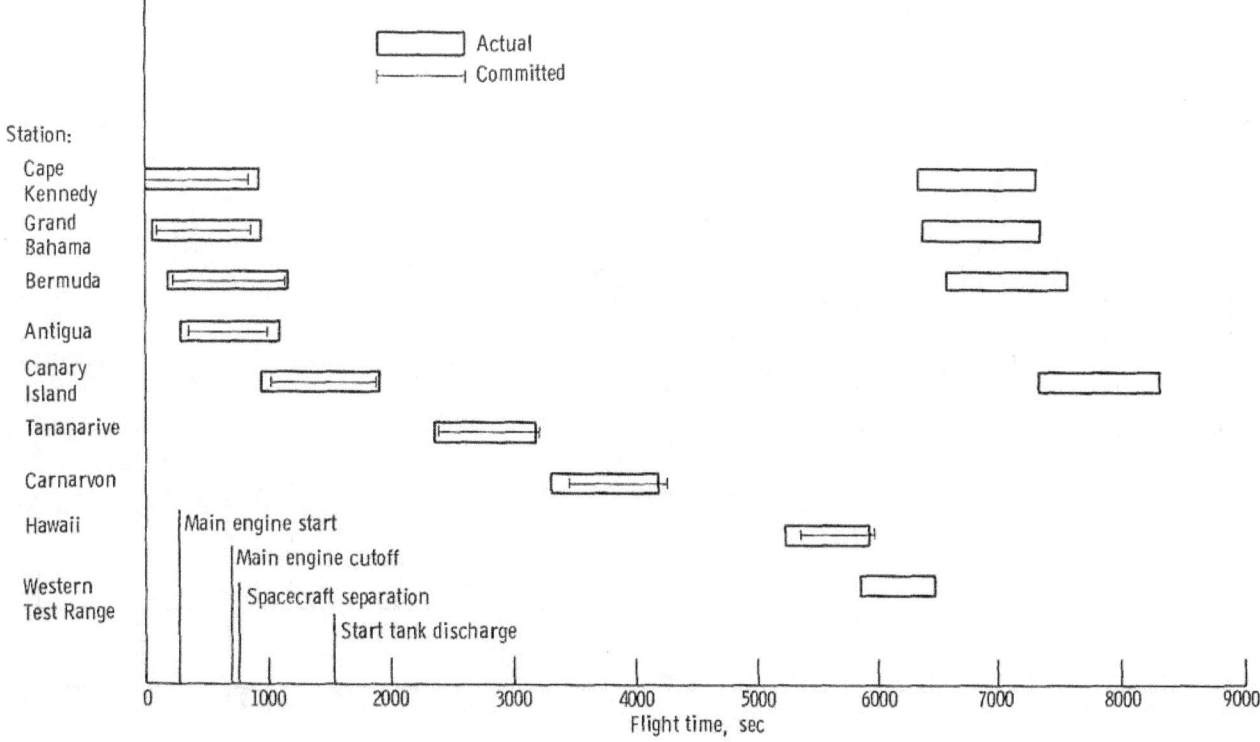

Figure VI-59. – Centaur telemetry coverage, AC-16.

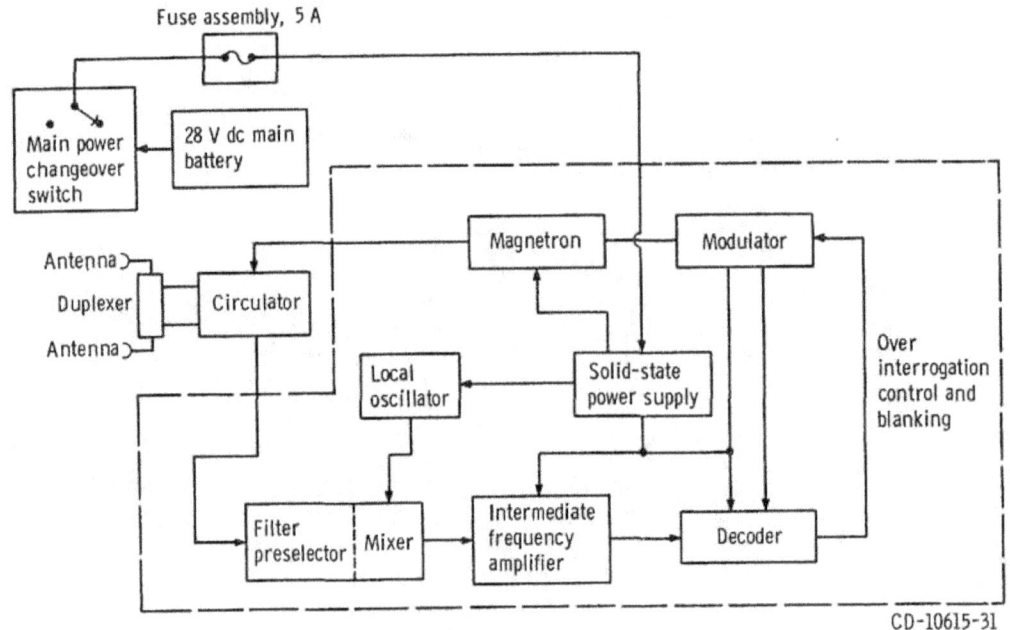

Figure VI-60. - Centaur C-band beacon subsystem, AC-16.

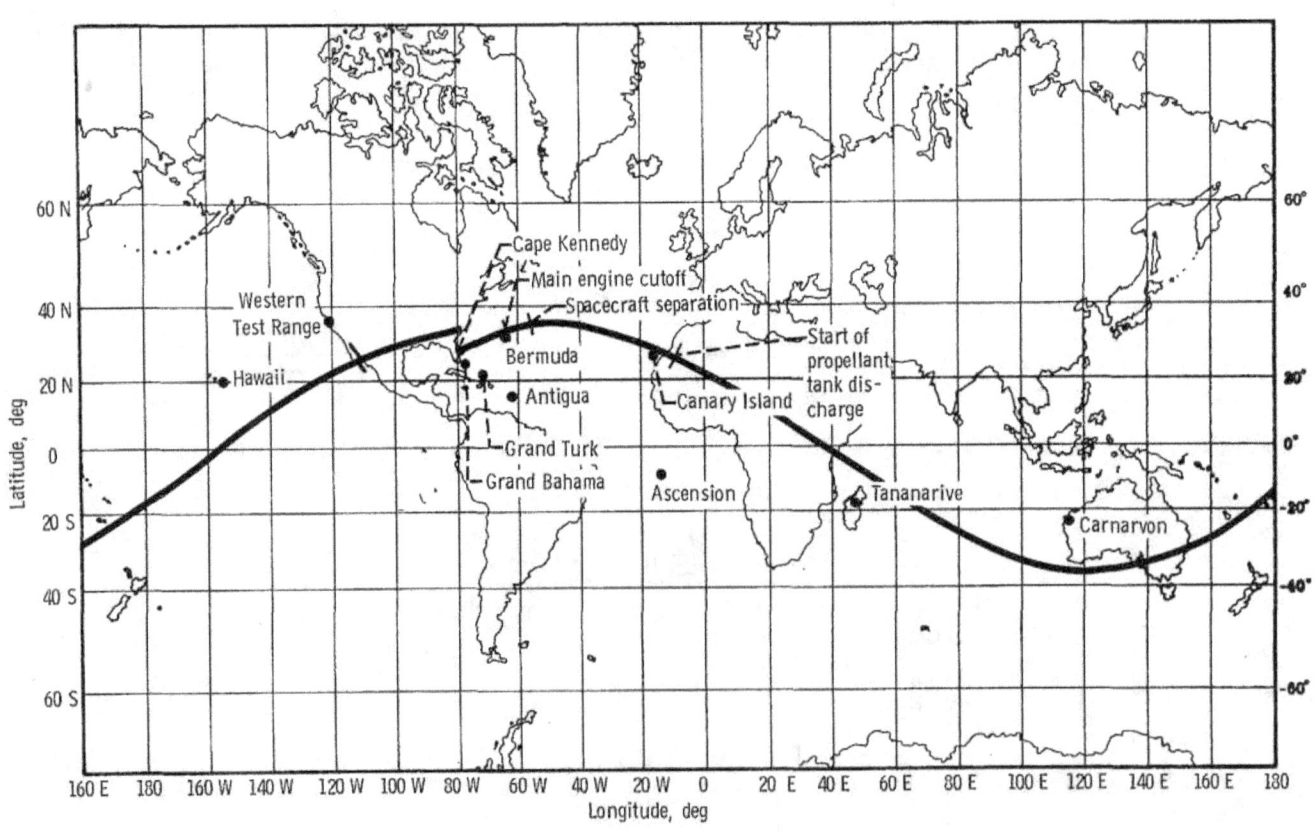

Figure VI-61. - Tracking station location and vehicle trajectory Earth track, AC-16.

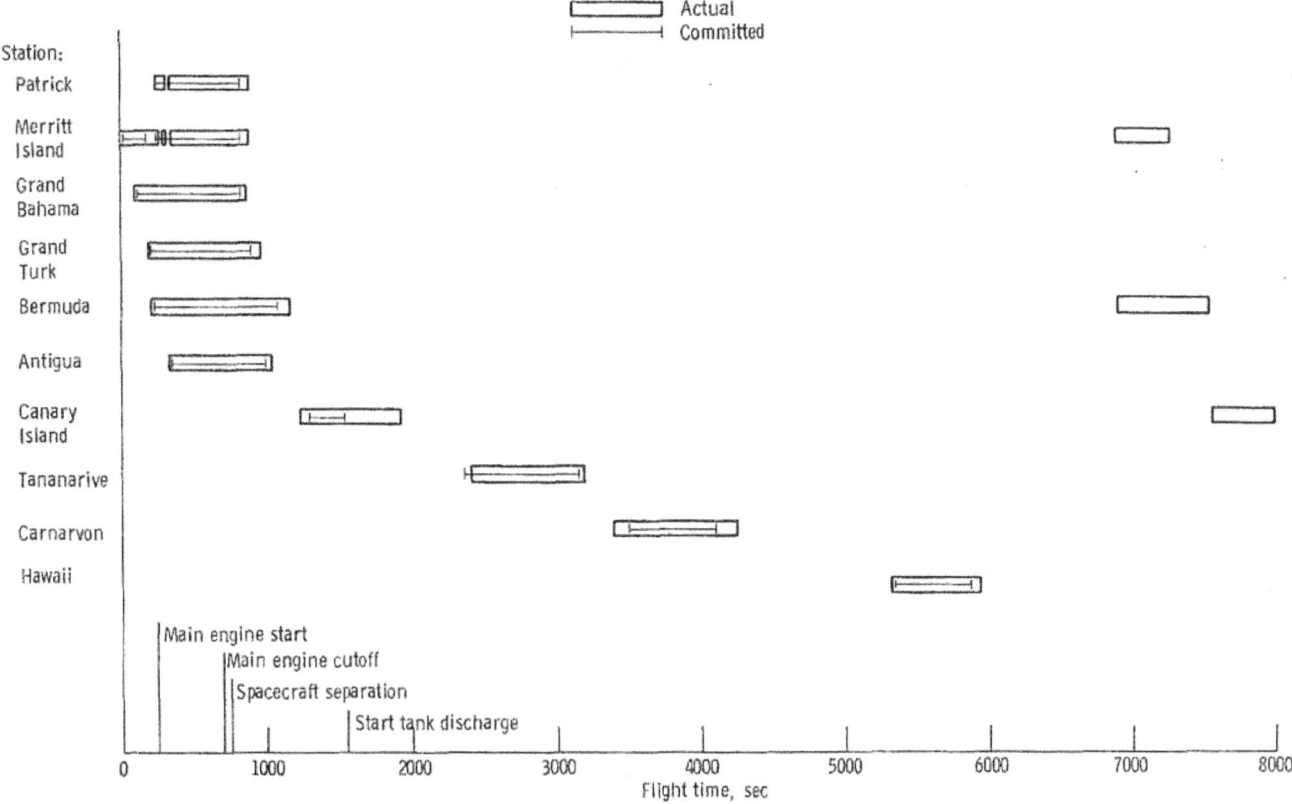

Figure VI-62. – C-Band radar coverage, AC-16. Coverage shown is for autobeacon track only.

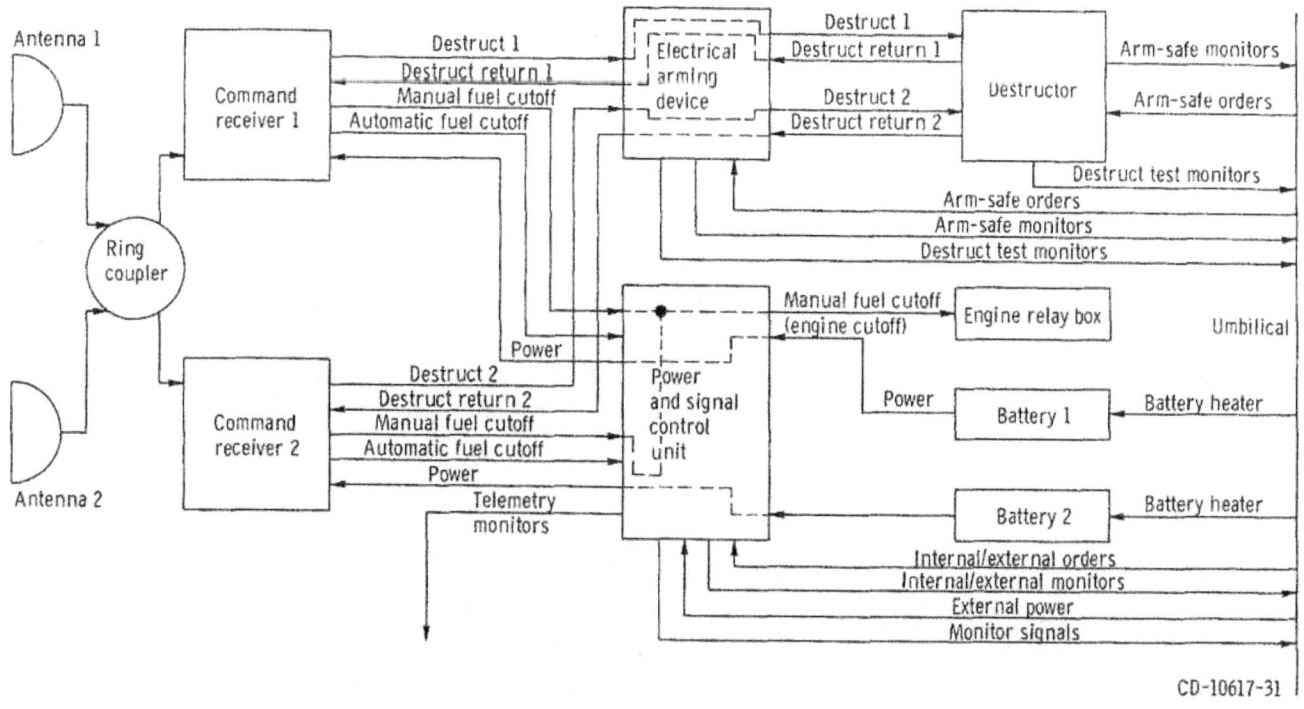

CD-10617-31

Figure VI-63. – Atlas vehicle destruct system block diagram, AC-16.

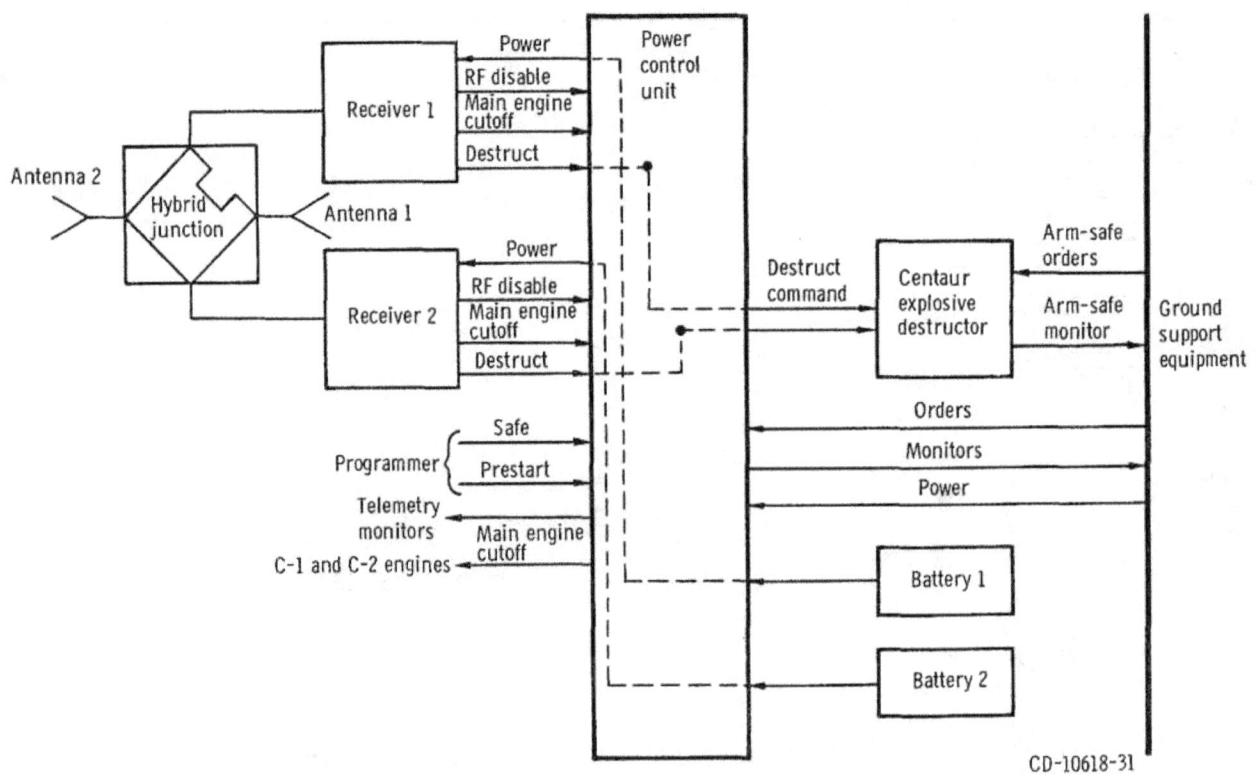

Figure VI-64. - Centaur vehicle destruct subsystem block diagram, AC-16.

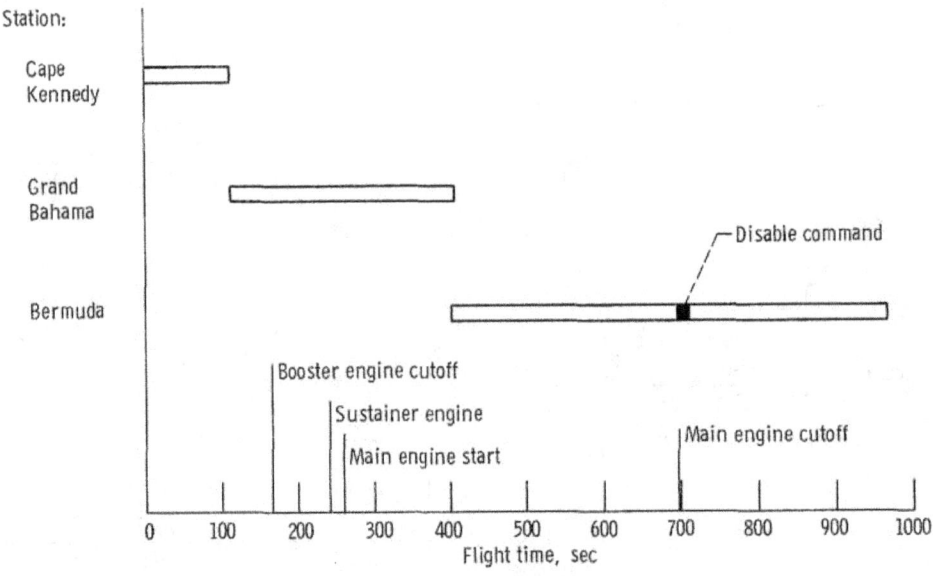

Figure VI-65. - Range safety command system transmitter utilization, AC-16.

GUIDANCE AND FLIGHT CONTROL SYSTEMS

by Paul W. Kuebeler, William J. Middendorf, and Corrine Rawlin

The objectives of the guidance and flight control systems are to guide the launch vehicle to the orbit injection point and establish the vehicle velocity necessary to place the spacecraft in the desired orbit. To accomplish these objectives, the guidance and flight control systems provide vehicle stabilization, control and guide the vehicle along the flight path, and sequence flight events of the launch vehicle. These functions are performed at specified time periods from vehicle lift-off through completion of the Centaur retromaneuver after spacecraft separation. An inertial guidance system is installed on the Centaur stage. Separate flight control systems are installed on the Atlas and on the Centaur stages. The guidance system, operating with the flight control systems, provides the capability to stabilize the vehicle and compensate for trajectory dispersions resulting from thrust misalinement, winds, and vehicle performance variations.

Three modes of operation are used for stabilization, control, and guidance of the launch vehicle. These modes are rate stabilization only, rate stabilization and attitude control, and rate stabilization and guidance control. Block diagrams of the three modes are shown in figure VI-66. The flight times during which a particular mode is used are shown in figure VI-67. Figure VI-67 also shows the modes of operation of the Centaur hydrogen peroxide attitude control system which are discussed later in this section.

The Atlas flight control system controls the Atlas-Centaur vehicle by gimbaling the Atlas main and vernier engines to provide thrust vector control. The Centaur flight control system controls the Centaur vehicle by gimbaling the main engines to provide thrust vector control while the main engines are firing or by commanding various combinations of the hydrogen peroxide attitude control engines on or off during coast periods, when the main engines are not firing.

The rate-stabilization-only mode stabilizes the roll axis of the Centaur stage continuously after Atlas-Centaur separation. This mode is also used to stabilize the pitch and yaw axes of the Centaur stage for 15.5 seconds following Atlas-Centaur separation. In this mode, output signals from rate gyros are used to control the vehicle. The output signal of each rate gyro is proportional to the angular rate of rotation of the vehicle about the input axis of the gyro. The vehicle angular rates are minimized by the flight control system to stabilize the vehicle. Rate stabilization is also combined with position (attitude) information in the other two modes of operation.

The rate-stabilization-and-attitude-control mode is used for pitch, yaw, and roll control during the Atlas booster phase of flight and for roll control only during the Atlas sustainer phase. This mode is termed attitude control since the displacement gyros (one each for the pitch, yaw, and roll axes) provide a reference attitude to which the vehicle is to be alined. However, if the actual flight path differs from the desired flight path, there is no way of determining the difference and correcting the flight path. The reference attitude is programmed to change during booster phase. These changes in reference attitude cause the vehicle to roll to the programmed flight azimuth angle and to pitch downward. Vehicle stabilization is accomplished in the same manner as in the rate-stabilization-only mode. The rate stabilization signals are algebraically summed with the attitude reference signals. These resultant signals are used to control and stabilize the vehicle.

The rate-stabilization-and-guidance-control mode is used for the pitch and yaw axes during Atlas sustainer phase, Centaur main engine firing, and the coast period following main engine cutoff. In this mode, the guidance system provides the attitude and direction reference. If the resultant flight path, as measured by the guidance system, is not the desired flight path, the guidance system issues steering signals to direct the vehicle to the desired flight path. Vehicle stabilization is accomplished in the same manner as in the rate stabilization only mode. The pitch and yaw rate stabilization signals are algebraically summed with the appropriate pitch and yaw steering signals from the guidance system. These resultant signals are used to control and stabilize the vehicle.

Figure VI-68 is a simplified diagram of the interface between the guidance system and the flight control system.

Guidance System

System description. - The guidance system performs the following functions:
(1) Measures vehicle acceleration in fixed inertial coordinates
(2) Computes the values of actual vehicle velocity and position, and computes the vehicle flight path to attain the trajectory injection point
(3) Compares the actual position to the desired flight path and issues steering signals
(4) Issues discrete commands

The Centaur guidance system is an inertial system which becomes completely independent of ground control approximately 12 seconds before lift-off of the vehicle. The guidance system consists of five separate units. Three of the units form an inertial measurement group. A simplified block diagram of the guidance system is shown in figure VI-69.

Inertial measurement group: Vehicle acceleration is measured by the following three units:
(1) The inertial platform unit, which contains the platform assembly, gyros, and accelerometers

(2) The pulse rebalance, gyro torquer, and power supply unit, which contains the electronics associated with the accelerometers

(3) The platform electronics unit, which contains the electronics associated with the gyros

The platform assembly uses four gimbals which provide a three-axis coordinate system. The use of four gimbals, instead of three, allows complete rotation of all three vehicle axes about the platform without gimbal lock. Gimbal lock is a condition in which two axes coincide, causing loss of one degree of freedom. A gimbal diagram is shown in figure VI-70. The azimuth gimbal is isolated from movements of the vehicle airframe by the other three gimbals. The inertial components (three gyros and three accelerometers) are mounted on the azimuth or inner gimbal. A gyro and an accelerometer are mounted as a pair with their sensing axes parallel. The gyro and accelerometer pairs are also alined on three mutually perpendicular (orthogonal) axes corresponding to the three axes of the platform.

The three gyros are identical and are of the single-degree-of-freedom, floated-gimbal, rate integrating type. Each gyro monitors one of the three axes of the platform. These gyros are elements of control loops, the sole purpose of which is to maintain each axis fixed in inertial space. The output signal of each gyro is connected to a servoamplifier whose output controls a direct-drive torque motor which moves a gimbal of the platform assembly. The orientation of the azimuth gimbal is fixed in the inertial space and the outer roll gimbal is attached to the vehicle. The angles between the gimbals provide a means for transforming steering signals from inertial coordinates to vehicle coordinates. The transformation is accomplished by electromechanical resolvers, mounted between gimbals, to produce analog electrical signals proportional to the sine and cosine functions of the gimbal angles. These electrical signals are used for an analog solution of the mathematical equations for coordinate transformation by interconnecting the resolvers in a multiple resolver chain.

The three accelerometers are identical and are of the single-axis, viscous-damped, hinged-pendulum type. The accelerometer associated with each axis measures the change in vehicle velocity along that axis by responding to acceleration. Acceleration of the vehicle causes the pendulum to move off center. The associated electronics then produce precise current pulses to recenter the pendulum. These rebalance pulses are either positive or negative depending on an increase or decrease in vehicle velocity. These pulses, representing changes in velocity (incremental velocity), are then routed to the navigation computer unit for computation of vehicle velocity.

Proper flight operation requires alinement and calibration of the inertial measuring unit during launch countdown. The azimuth of the platform, to which the desired flight trajectory is referenced, is aligned by ground-based optical equipment. The platform is alined perpendicular to the local vertical by using the two accelerometers in the horizontal plane. Each gyro is calibrated to determine its characteristic constant torque drift rate and mass unbalance along the input axis. The scale factor and zero bias off-set of each accelerometer is determined. These prelaunch-determined calibration constants and scale factors are stored in the navigation computer for use during flight.

Navigation computer unit: The navigation computer unit is a serial, binary, digital machine with a magnetic drum memory. The memory drum has a capacity of 2816 words (25 bits/word) of permanent storage, 256 words of temporary storage, and six special-purpose tracks. Permanent storage is prerecorded and cannot be altered by the computer. The temporary storage is the working storage of the computer.

Incremental velocity pulses from the accelerometers are the information inputs to the navigation computer. The operation of the navigation computer is controlled by the prerecorded program. This program directs the computer to use the prelaunch equations, navigation equations, and guidance equations.

The prelaunch equations establish the initial conditions for the navigation and guidance equations. Initial conditions include (1) a reference trajectory, (2) launch site values of geographical position, and (3) initial values of navigation and guidance functions. Based on these initial conditions, the guidance system starts flight operation approximately 12 seconds before lift-off.

The navigation equations are used to compute vehicle velocity and actual position. Velocity is determined by algebraically summing the incremental velocity pulses from the accelerometers. An integration is then performed on the computed velocity to determine actual position. Corrections for the prelaunch-determined gyro and accelerometer constants are also made during the velocity and position computation to improve the navigation accuracy. For example, the velocity data derived from the accelerometer measurements are adjusted to compensate for the accelerometer scale factors and zero offset biases measured during the launch countdown. The direction of the velocity vector is also adjusted to compensate for the gyro constant torque drift rates measured during the launch countdown.

The guidance equations continually compare actual position and velocity with the position and velocity desired at the time of injection. Based upon this position comparison, steering signals are generated to guide the vehicle along an optimized flight path to obtain the desired injection conditions. The guidance equations are used to generate four discrete commands: (1) booster engine cutoff, (2) sustainer engine cutoff backup, (3) Centaur main engine cutoff, and (4)"null" the propellant utilization system. The booster engine cutoff command and the sustainer engine cutoff backup commands are issued when the measured vehicle acceleration equals predetermined values. The Centaur main engine cutoff command is issued when the extrapolated vehicle orbital angular momentum equals that required for injection into orbit. The command to "null" the propellant utilization system is issued 15 seconds before the Centaur main engine cutoff command.

During the booster phase of flight, the navigation computer supplies incremental pitch and yaw signals for steering the Atlas stage. From a series of predetermined programs, one pitch program and one yaw program are selected based on prelaunch upper-air wind soundings. The selected programs are entered and stored in the computer during launch countdown. The program consists of discrete pitch and yaw turning rates for specified time intervals from $T + 15$ seconds until booster engine cutoff. These programs permit changes to be made in the flight reference trajectory during countdown to reduce anticipated aerodynamic heating and structural loading conditions on the vehicle.

Signal conditioner unit: The signal conditioner unit is the link between the guidance system and the vehicle telemetry system. This unit modifies and scales guidance system parameters to match the input range of the telemetry system.

System performance. - The performance of the guidance system was as follows:

Accuracy: The performance of the guidance system was satisfactory, as indicated by the following orbital parameters:

| Parameter | Units | Predicted value | Actual value | | Error, predicted value minus BET value | Estimated 3-sigma errors |
			GRT[a]	BET[b]		
Injection[c]	km	772.676	772.716	772.768	0.092	1.46
	n mi	416.930	416.952	416.980	.050	.79
Apogee minus	km	1.275	1.188	3.154	1.879	10.23
perigee	n mi	.688	.592	1.702	1.014	5.52
Eccentricity	----	.000089	.000083	.00022	-----	------
Inclination	deg	34.9982	34.9990	34.9811	0.017	0.035

[a]Guidance Reconstructed Trajectory obtained from telemetered guidance data.
[b]Best Estimated Trajectory obtained from radar tracking data.
[c]See V. TRAJECTORY AND PERFORMANCE for details on injection altitude.

Discrete commands: All discrete commands were issued properly. Table VI-XI lists the discretes, the criteria for the issuance of the discretes, and the computed values at the time the discretes were issued. Actual and predicted times from lift-off are also shown for reference only.

Guidance steering loop: The pitch and yaw steering signals issued by the guidance system are proportional to the components of the steering vector (desired vehicle pointing vector) along the vehicle pitch and yaw axes. In this section of the report, the steering signals have been converted into the approximate angular attitude errors between the steering vector and the vehicle roll axis (vehicle pointing vector) in the pitch and yaw planes. During the sustainer phase of flight and during Centaur main engine firing, the attitude errors remained less than 2.5^O in pitch and 1^O in yaw, except during periods when guidance steering was deactivated or when a large sudden change in the steering vector occurred.

Attitude errors at guidance steering activation are given in the following table:

Event	Time, sec	Attitude displacements, deg	
		Pitch	Yaw
Sustainer phase	T + 160	1 (nose below)	<1 (nose right)
Centaur powered flight	T + 250	1.5 (nose below)	2.2 (nose right)

At T + 267.1 seconds the guidance system issued a nose right command of approximately 14° in the inertial coordinate system. The resultant command in the vehicle coordinate system was approximately 13.8^O nose right and 4.5^O nose down. The retromaneuver command was issued at T + 1055.2 seconds, and the vehicle began to aline to approximately the negative of the instantaneous radius vector.

Accelerometer loops and gyro control loops: The accelerometer loops operated satisfactorily. The accelerometer pendulum offsets from "null" remained within a band of approximately 2 arc-seconds.

The gimbal control loops operated satisfactorily, and the inertial platform remained stable throughout the flight. The fourth gimbal uncaged at T + 82 seconds, when the vehicle had pitched over approximately 23° as planned. The maximum gimbal displacement errors are given in the following table:

Gimbal	Maximum displacement errors, arc-sec	
	Through main engine cutoff	After main engine cutoff
1	7.8, [a]-8.0	7.8, -7.8
2	7.4, -8.0	5.5, -7.3
3	11.6, -7.6	12.0, -11.6
4	[a]335, -301	417, -668

[a]Excludes gimbal 4 uncaging transient of -12.9 arc-sec for gimbal 1 and 2600 arc-sec for gimbal 4.

Other measurements: Except for the digital data, all of the guidance system signals and measurements which were monitored during the flight were normal and indicated satisfactory operation of the guidance system. The skin temperature of the Pulse Rebalance Electronics was 288.2 K (59^{0} F) at lift-off and reached 302.1 K (84^{0} F) at the end of the first orbital pass.

After T + 92 minutes, an intermittent navigation computer malfunction resulted in incorrect digital telemetry data. Incorrect bits occurred, intermittently, during the telemetry work marker and the data words and following some of the data words. The malfunction affected only the telemetry output function of the computer. Otherwise, the computer performed normally. It is believed that the malfunction was the intermittent failure of the telemetry output flip-flop (located in the computer) to reset upon command, because of the temperature increase of the flip-flop during its extended operation. This specific malfunction has been reproduced in laboratory tests.

Flight Control Systems

System description - Atlas. - The Atlas flight control system provides the primary functions required for vehicle stabilization, control, sequencing, and execution of guidance steering signals, and consists of the following major units:

(1) The displacement gyro unit, which contains three single-degree-of-freedom, floated, rate integrating gyros and associated electronic circuitry for gain selection and signal amplification. These gyros are mounted to the vehicle airframe in an orthogonal triad configuration alining the input axis of a gyro to its respective vehicle axis of pitch, yaw, or roll. Each gyro provides an electrical output signal proportional to the integral of the time rate of change of angular displacement from the gyro reference axis.

(2) The rate gyro unit, which contains three single-degree-of-freedom, floated, rate gyros and associated electronic circuitry. These gyros are mounted in the same manner as the displacement gyro unit. Each gyro provides an electrical output signal proportional to the angular rate of rotation of the vehicle about the gyro input (reference) axis. The AC-16 flight was the first wherein the Centaur provided both rate and position signals to the Atlas flight control system during the Atlas sustainer phase.

(3) The servoamplifier unit, which contains electronic circuitry to amplify, filter, integrate, and algebraically sum combined position and rate signals with engine position feedback signals. The electrical outputs of this unit direct the hydraulic actuators which gimbal the engines to provide thrust vector control.

(4) The programmer unit, which contains an electronic timer, arm-safe switch, high-, low-, and medium-power electronic switches, and circuitry to set the roll program from launch ground equipment. The programmer issues discrete commands to the following systems: other units of the Atlas flight control system, Atlas propulsion, Atlas pneumatic, vehicle separation systems, and Centaur flight control.

System performance - Atlas. - The flight control system performed satisfactorily throughout the Atlas phase of flight. Corrections required to control the vehicle because of disturbances were well within the system capability. Vehicle dynamic response resulting from each flight event was evaluated in terms of amplitude, frequency, and duration as observed on rate gyro data (table VI-XII). In this table, the control capability (in percent) is the ratio of engine gimbal angle used to the available total engine gimbal angle.

The programmer was started at 1.1-meter (42-in.) rise, which occurred at approximately T + 1 second. This permitted the flight control system to gimbal the engines and thereafter control the vehicle. The vehicle lift-off transients were damped out by T + 4 seconds using 12 percent of the control capability.

At T + 2 seconds the roll program was initiated and continued until T + 20 seconds; this was 5 seconds longer than for previous flights. This large roll maneuver was required for this flight to satisfy the Bermuda overflight constraints. (See V. TRAJECTORY AND PERFORMANCE for discussion of Bermuda overflight constraints.) Atlas and Centaur rate gyro data indicated that the vehicle rolled clockwise approximately 53°. This value compares favorably with an expected roll of 55° required to achieve the desired flight azimuth of 60°.

The pitch program was started at T + 15 seconds with a pitch rate of -0.36 degree per second.

Rate gyro data indicated that the period of maximum aerodynamic loading for this flight was approximately from T + 75 to T + 95 seconds. During this period, a maximum of 48 percent of the control capability was required to overcome both steady-state and transient loading.

Atlas booster engine cutoff occurred at T + 152.1 seconds. The angular rates imparted to the vehicle by this transient required 8 percent of the sustainer engine gimbal capability. The Atlas booster engines were jettisoned at T + 155.2 seconds. The maximum angular rate imparted by this disturbance was 3.11 degree per second in the yaw plane.

During the Atlas powered flight, the Atlas flight control system provided the attitude reference until T + 160.2 seconds; thereafter, the Centaur guidance system provided the attitude reference. Twenty percent of the total control capability was required to steer the vehicle to the new reference. The maximum vehicle angular rate transient during this change was a yaw rate of 1.64 degree per second, peak-to-peak, with a duration of 7 seconds.

Insulation panels were jettisoned at T + 196.8 seconds. The maximum vehicle angular rate transient observed due to this disturbance was a pitch rate of 2.32 degree per second, peak-to-peak, which utilized 20 percent of the control capability for correction.

Sustainer engine cutoff occurred at $T + 234.5$ seconds. Atlas-Centaur separation was smooth and resulting transients were under 0.5 degree per second.

System description - Centaur. - The Centaur flight control system provides the primary means for vehicle stabilization and control, execution of guidance steering signals, and timed switching sequences for programmed flight events. The Centaur flight control system (fig. VI-71) consists of the following major units:

(1) The rate gyro unit, which contains three single-degree-of-freedom, floated, rate gyros with associated electronics for signal amplification gain selection and conditioning of guidance steering signals. These gyros are mounted to the vehicle in an orthogonal triad configuration alining the input axis of each gyro to its respective vehicle axis of pitch, yaw, or roll. Each gyro provides an electrical output signal proportional to the angular rate of rotation of the vehicle about the gyro input (reference) axis.

(2) The servoamplifier unit, which contains electronics to amplify, filter, integrate, and algebraically sum combined position and rate signals with engine position feedback signals. The electrical outputs of this unit issue signals to the hydraulic actuators which control the gimbaling of the engines. In addition, this unit contains the logic and threshold circuitry controlling the engines of the hydrogen peroxide attitude control system.

(3) The electromechanical sequence timer unit, which contains a 400-hertz synchronous motor to provide the time reference and 22 time slots capable of actuating 144 switches.

(4) The auxiliary electronics unit which contains logic, relay switches, transistor switches, power supplies, control circuitry for the electromechanical timer, circuitry for conditioning computer generated discretes, and an arm-safe switch. The arm-safe switch electrically isolates valves and pyrotechnic devices from the control switches. The combination of the electromechanical timer units and the auxiliary electronics unit issues discretes to the following systems: other units of the Centaur flight control, propulsion, pneumatic, hydraulic, separation, propellant utilization, telemetry, spacecraft, and electrical systems, and the Atlas flight control system.

Vehicle steering during Centaur powered flight is by thrust vector control through gimbaling of the two main engines. There are two actuators for each engine to provide pitch, yaw, and roll control. Pitch control is accomplished by moving both engines together in the pitch plane. Yaw control is accomplished by moving both engines together in the yaw plane, and roll control is accomplished by moving the engines differentially in the yaw plane. Thus, the yaw actuator responds to an algebraically summed yaw-roll command. By controlling the direction of thrust of the main engines, the flight control system maintains the flight of the vehicle on a trajectory directed by the guidance system. After main engine cutoff, control of the vehicle is maintained by the flight control system through selective firing of hydrogen peroxide engines. A more complete description of the engines and the propellant supply for the attitude control system is presented in the section PROPULSION SYSTEMS.

The logic circuitry, which commands the 14 hydrogen peroxide engines either on or off, is contained in the servoamplifier unit of the flight control system. Figure VI-72 shows the alphanumeric designations of the engines and their locations on the aft end of the vehicle. Algebraically summed position and rate signals are the inputs to the logic circuitry. The logic circuitry provides five modes of operation designated: all off, separate on, A and P separate on, V half on, and S half on. These modes of operation are used during different periods of the flight and are controlled by the sequence timer unit. A summary of the modes of operation is presented in table VI-XIII. In this table "threshold" designates the vehicle rate in degrees per second that has to be exceeded before the engines are commanded "on."

System performance - Centaur. - The Centaur flight control system performance was satisfactory. Vehicle stabilization and control were maintained throughout the flight. All events sequenced by the timer were executed at the required times. (Refer to fig. VI-67 for the time periods of guidance - flight control mode of operation and attitude control system mode of operation. Vehicle dynamic responses for selected flight events are tabulated in table VI-XII.) The following evaluation of system performance is presented in the order of time sequenced portions of the flight.

Sustainer engine cutoff to Centaur main engine cutoff (T + 234.5 to T + 698.2 sec): The Centaur timer was started at Atlas sustainer engine cutoff by a command from the Atlas programmer. Appropriate commands were issued to separate the Centaur from the Atlas and to initiate the Centaur main engine firing sequence. Vehicle control was maintained during the period between sustainer engine cutoff and main engine start by gimbaling the main engines as they were discharging boost pump turbine exhaust and chilldown flow. There were no significant transients during separation. Centaur main engine start occurred at T + 246.0 seconds; only small transients were observed. Four seconds after main engine start the steering vector generated at sustainer engine cutoff was admitted and held until T + 267.1 seconds, 9.5 seconds after nose fairing jettison; this prevented the vehicle from turning during this time interval in order to minimize vehicle motion during the nose fairing jettison event. Nose fairing jettison occurred at T + 257.6 seconds with maximum response in roll, as expected. Nose fairing jettison was programmed to occur after Atlas-Centaur separation. At T + 267.1 seconds a yaw right maneuver was performed to achieve the required orbit inclination of 35°. The yaw right maneuver (dogleg) introduced a 2.94-degree-per-second yaw rate, which was damped in 7 seconds. Vehicle angular rates during the remainder of Centaur burn were larger than for previous flights. Angular rates as high as 1.66 degrees per second were observed during this time. This was partly due to the fact that the guidance compute cycle for AC-16 and AC-17 was on the order of three times larger than for previous Atlas-

Centaur flights. The long compute cycle caused the vehicle to steer along one vector for approximately 4.5 seconds while a new vector was being computed and commanded. The powered phase transients on AC-17 with a comparable compute cycle were smaller than for AC-16 because of the effect of larger end-to-end position gains for the AC-16 autopilot. At all times the transients were well within the vehicle control capability.

Centaur main engine cutoff to retrothrust (T + 698.2 to T + 1548.2 sec): The hydrogen peroxide attitude control system was activated at the time of main engine cutoff in the A and P separate-on mode. Angular rates imparted to the vehicle due to the differences in the shutdown characteristics of the two Centaur main engines were reduced to acceptable control levels in 1 second. During the time from main engine cutoff to spacecraft separation, the vehicle alined to the separation vector. Also during this time interval the spacecraft solar paddles were deployed, and the balance booms were extended. These events were initiated at T + 708.2 and T + 723.2 seconds, respectively. The only significant angular rate due to these disturbances was a roll rate of 1.69 degrees per second due to solar paddle deployment. Only small angular rates were observed when the spacecraft was separated from the Centaur at T + 748.3 seconds.

At T + 1055.2 seconds the Centaur began its reorientation to aline its axis along the approximately negative geocentric radius vector. The purposes of the retromaneuver were to minimize the impingement of Centaur exhaust on the spacecraft and to provide a difference between the orbital periods of the Centaur and the OAO spacecraft. The attitude control system then switched to the V half-on mode for 50 seconds. This was followed by the S half-on mode which lasted until the initiation of retrothrust. The attitude control system sequence was so designed because of the viewing constraints (cone angle) imposed upon the Centaur by the spacecraft. (See V. TRAJECTORY AND PERFORM-ANCE for discussion on viewing constraints (cone angles).) Unbalanced venting of hydrogen began during the S half-on mode. Venting started at approximately T + 1493 seconds, which was 58 seconds prior to "blowdown." Rate gyro data indicated that the V engines came on to control about 1.5 seconds after venting began. A peak disturbance torque of approximately 289 newton-meters (2560 in.-lbf) was calculated from yaw rate gyro data. The disturbing torque exceeded the control torque for about 3 seconds. The maximum angular rate due to this disturbance was approximately 0.58 degree per second.

At about T + 1548 seconds the engine prestart valves were opened to allow the residual propellants to discharge through the main engines.

TABLE VI-XI. - DISCRETE COMMANDS, AC-16

Discrete command	Criteria for discrete to be issued	Discrete issued at this computed value	Actual time, T + sec	Predicted time, T + sec
Booster engine cutoff	When square of vehicle thrust acceleration is greater than 27.53 $(g's)^2$ [a]$(5.25 g's)^2$	31.92 $(g's)^2$ $(5.65 g's)^2$	152.1	153
Sustainer engine cutoff backup	When square of vehicle thrust acceleration is less than 0.53 $(g's)^2$	0.36 $(g's)^2$	238	240
Null propellant utilization system	Fifteen seconds before main engine cutoff discrete	16.7 sec	681.5	671

[a]This value is lower than the vehicle thrust acceleration of 5.7 ± 0.113 g required for issuing the discrete to allow for time delays between the time the computer determines the criteria is satisfied until the discrete is actually issued.

TABLE VI-XII. – VEHICLE DYNAMIC RESPONSE TO FLIGHT DISTURBANCES

Event	Flight time, T + sec	Measurement	Rate gyro peak-to-peak amplitude, deg/sec	Transient frequency, Hz	Transient duration, sec	Required control capability, percent
Lift-off	0	Pitch	0.50	(b)	----	(c)
		Yaw	.49	(b)	----	(c)
		Roll	.48	(b)	----	(c)
42-inch (1.1-m) rise	1	Pitch	3.31	5	4	12
		Yaw	.49	----	----	4
		Roll	1.77	1.43	1	4
Region of maximum aerodynamic pressure	82 / 87 / 83	Pitch	1.98	0.20	(d)	44
		Yaw	1.47	.22	(d)	48
		Roll	1.12	.42	(d)	48
Booster engine cutoff	152.2	Pitch	1.32	10	(e)	8
		Yaw	1.14	10	(e)	4
		Roll	1.28	10	(e)	8
Booster engine jettison	155.1	Pitch } High frequency	1.66	3.3	10	(f)
		Yaw }	2.29	3.3	12	(f)
		Roll }	2.41	20	1	(f)
		Pitch } Low frequency	1.49	.67	10	(f)
		Yaw }	3.11	.72	12	(f)
		Roll }	1.12	.63	12	(f)
Admit guidance	160.2	Pitch	1.32	0.35	5	8
		ᵃYaw	1.64	.20	7	20
		Roll	(b)	(b)	(h)	8
Insulation panel jettison	196.8	Pitch	2.32	25	0.6	20
		Yaw	.98	25	.6	6
		Roll	.40	25	1	6
			1.12	.2	1	6
Sustainer engine cutoff	233.4	Pitch	0.50	20	1.5	25
		Yaw	.49	20	1.5	20
		Roll	(i)	(i)	----	8
Atlas-Centaur separation	236.6	Pitch	0.33	-----	0.3	--
		Yaw	.49	-----	.3	--
		Roll	(i)	(i)	----	--
Main engine start	247.2	Pitch	0.66	20	2.9	18
			.99	4	2.9	18
			.65	20	2.9	24
			.64	.5	1.0	24
Admit guidance	250.1	Pitch	0.66	0.5	7.5	26
		Yaw	1.14	.5	5.5	46
		Roll	(i)	(i)	----	--
Nose fairing jettison	258	Pitch	1.16	25	1	4
		Yaw	1.06	25	1.4	12
		Roll	1.76	2	2	12
Transient due to holding attitude for nose fairing jettison	267.1	Pitch	1.96	(j)	7	70
		Yaw	2.94	(j)	7	76
		Roll	.32	(j)	7	76
Main engine cutoff	698.3	Pitch	3.81	30	1	8
		Yaw	3.76	30	1	8
		Roll	1.28	30	5	8
Deploy solar paddles	708.3 / 711.3	Pitch	(k)	(k)	----	--
		Yaw	(k)	(k)	----	--
		Roll	1.69	-----	6	--
Extend balance booms	726.3	Pitch	(k)	(k)	----	--
		Yaw	0.25	1.43	5	--
		Roll	.16	1.43	10	--
Spacecraft separation	748.3	Pitch	0.50	40	3	--
		Yaw	.65	25	3	--
		Roll	.48	1.25	6	--

ᵃTime of transient as indicated on rate gyro data.
ᵇToo varied to measure.
ᶜAutopilot not yet active.
ᵈTransients during period from T + 75 to T + 95 sec.
ᵉNearly damped by time of booster engine jettison.
ᶠSustainer engine control inactive during booster engine jettison.
ᵍAppears to be damped rate from booster engine jettison.
ʰOnly damped rate from booster engine jettison remains.
ⁱNo measurable transient.
ʲDifficult to determine - odd shape.
ᵏToo small to measure.

137

TABLE VI-XIII. - DESCRIPTION OF ATTITUDE CONTROL SYSTEM MODES OF OPERATION, AC-16

[V engines, 222.4 N (50 lbf) thrust; S engines, 13.3 N (3.0 lbf) thrust; A engines, 15.6 N (3.5 lbf) thrust; P engines, 26.7 N (6.0 lbf) thrust.]

Mode	Flight period	Description
All off	Powered phases	This mode inhibits the operation of all attitude control engines.
Separate on	From 850 sec after main engine cutoff until retrothrust	When in the separate-on mode, a maximum of two V and two A engines and one P engine fire. These engines fire only when appropriate error signals surpass their respective threshold. A engines: When 0.2-deg/sec threshold is exceeded, suitable A engines fire to control in yaw and roll. A_1A_4 and A_2A_3 combinations are inhibited. P engines: When 0.2-deg/sec threshold is exceeded, suitable P engine fires to control in pitch. P_1P_2 combination is inhibited. S engines: Off V engines: When 0.3-deg/sec threshold is exceeded, suitable V engines fire (as a backup for higher rates). V_1V_3 and V_2V_4 combinations are inhibited.
A and P separate on	For 451 sec after main engine cutoff	This mode is the same as separate-on mode, except V engines are inhibited.
V half on	From 451 until 500 sec after main engine cutoff	A engines: When 0.2-deg/sec threshold is exceeded, suitable A engines fire to control in roll only. P engines: Off S engines: Off V engines: When there are no error signals, V_2V_4 combination fires continuously. The continuous firing provides lateral and added longitudinal separation between Centaur and spacecraft. When 0.2-deg/sec threshold is exceeded, a minimum of two and a maximum of three V engines fire to control in pitch and yaw.
S half on	From 500 until 850 sec after main engine cutoff	A engines: When 0.2-deg/sec threshold is exceeded, suitable A engines fire to control in roll only. P engines: Off S engines: When there are no error signals, S_2S_4 combination fires continuously. When 0.2-deg/sec threshold is exceeded, a minimum of two and a maximum of three engines fire to control in pitch and yaw. V engines: When 0.3-deg/sec threshold is exceeded, a minimum of one and a maximum of two V engines fire to control in pitch and yaw. When a V engine fires, the corresponding S engine is commanded off.

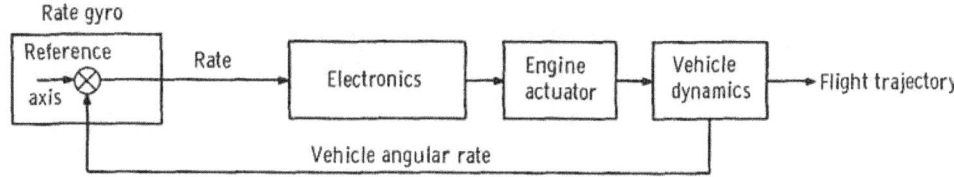

(a) Rate-stabilization-only mode.

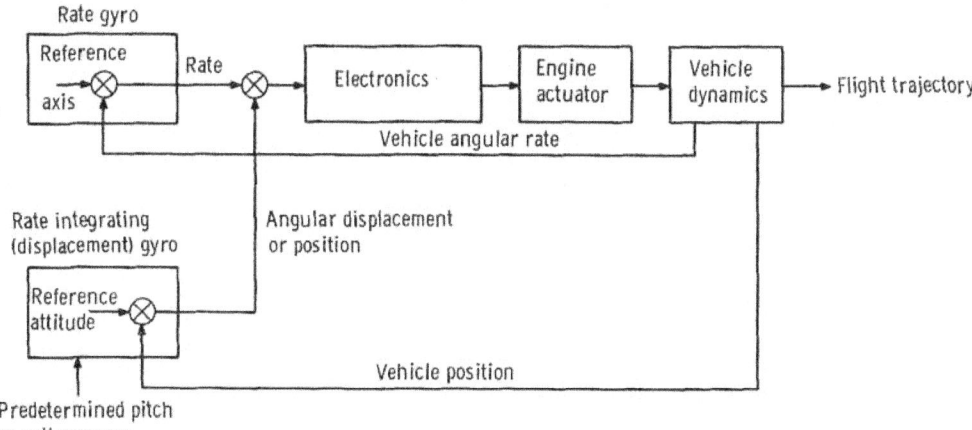

(b) Rate-stabilization-and-attitude-control mode.

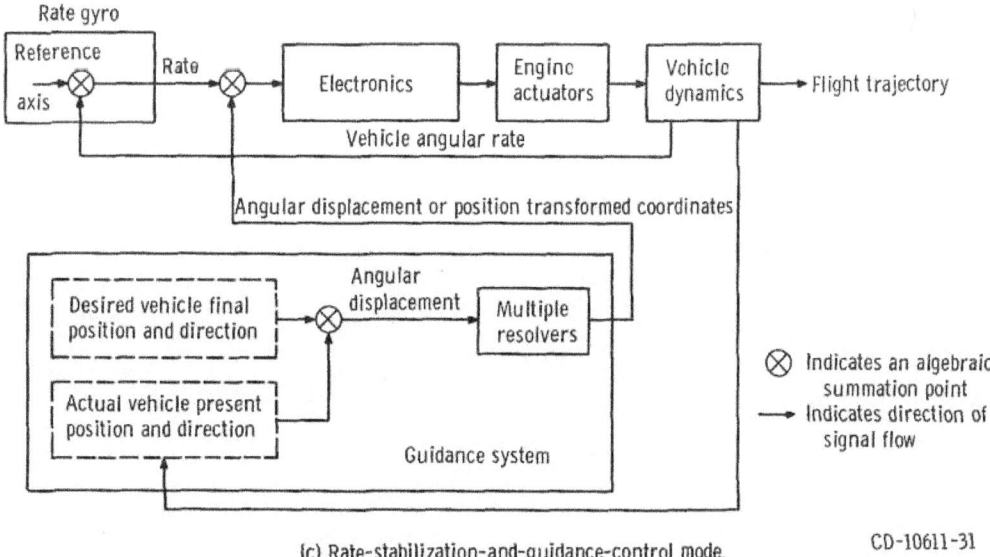

(c) Rate-stabilization-and-guidance-control mode.

CD-10611-31

Figure VI-66. - Guidance and flight control modes of operation, AC-16.

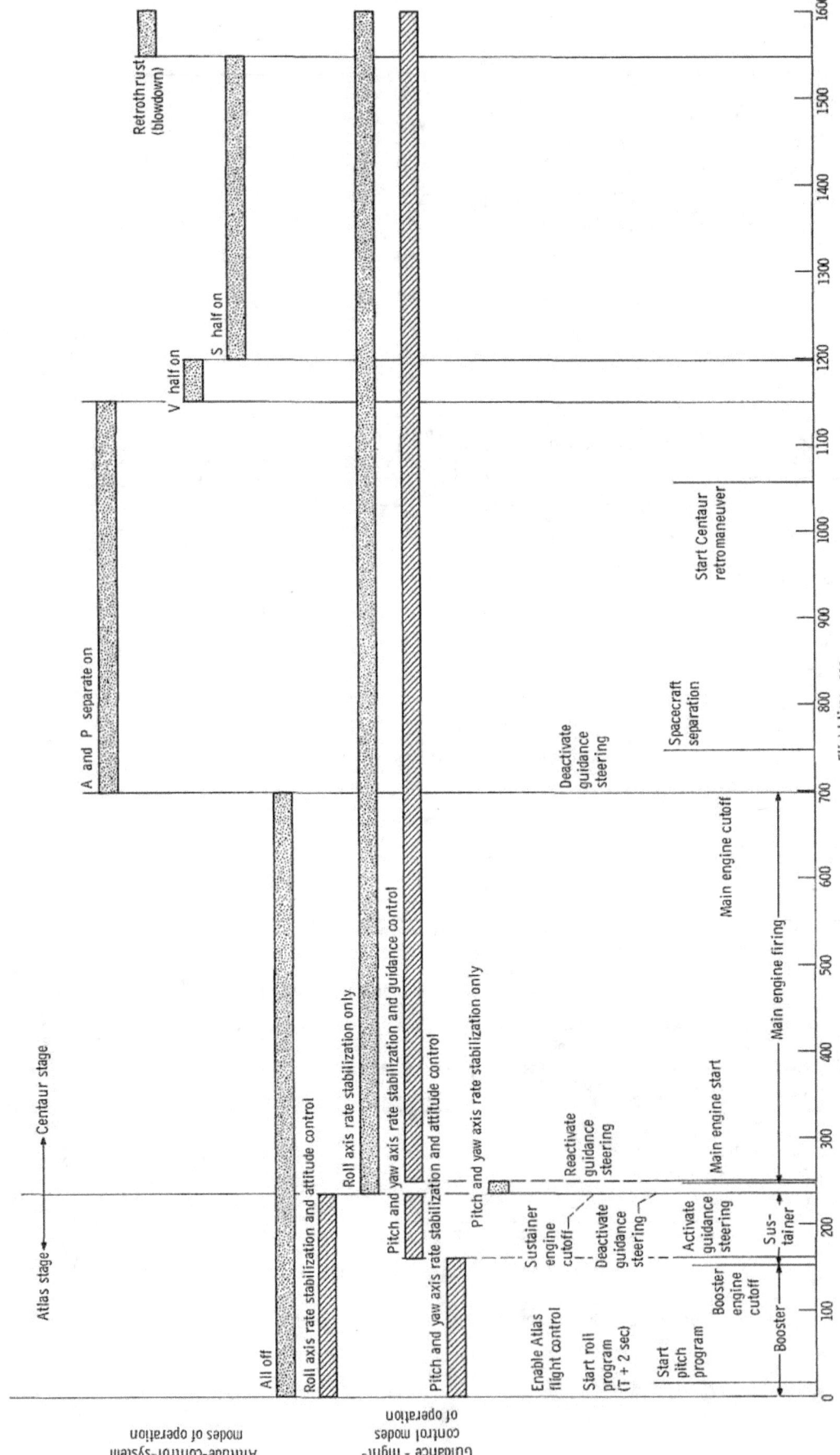

Figure VI-67. – Time periods of guidance – flight-control modes and attitude-control-system modes of operation. (There is no rate control during a 3.5-second period following sustainer engine cutoff because the engines are not developing thrust.)

140

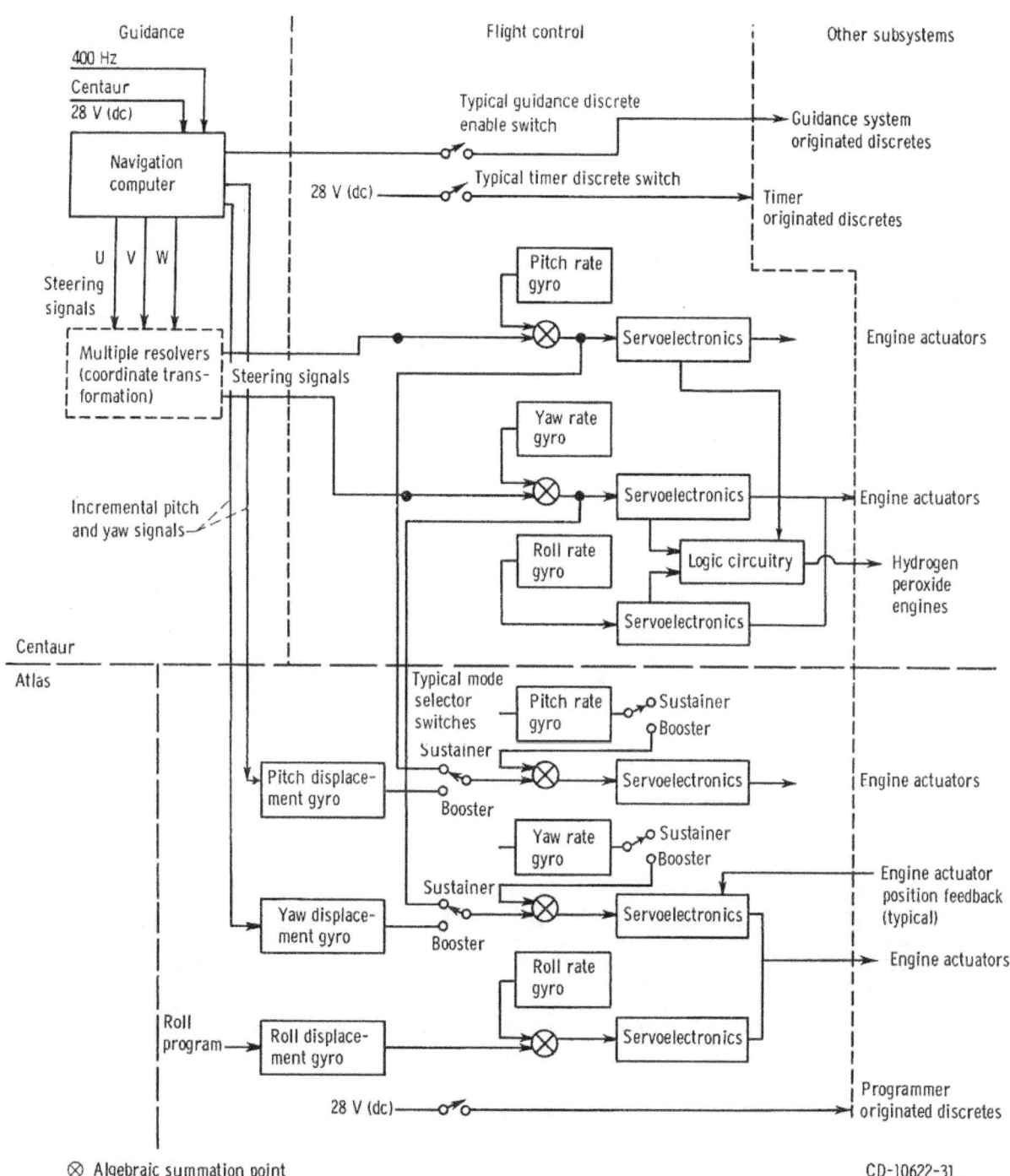

⊗ Algebraic summation point

Figure VI-68. - Simplified guidance and flight control systems interface, AC-16.

CD-10622-31

141

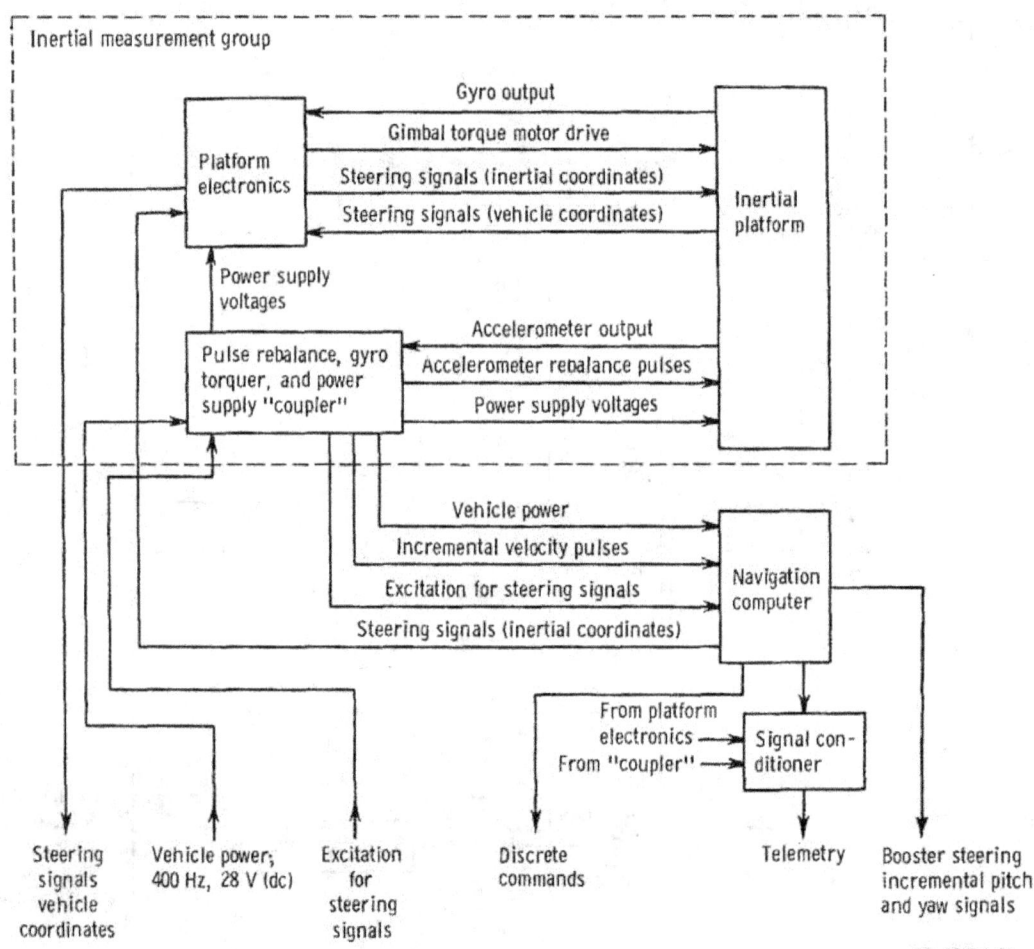

Figure VI-69. - Simplified block diagram of Centaur guidance system, AC-16.

CD-10616-31

142

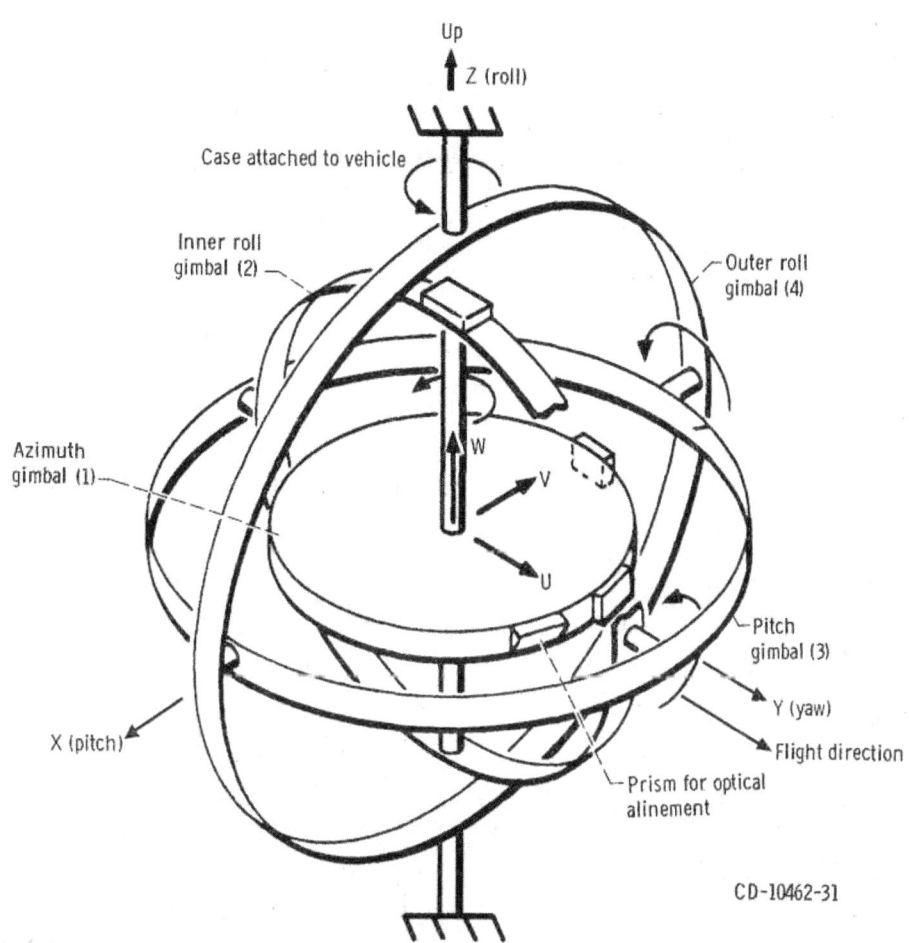

Up

Z (roll)

Case attached to vehicle

Inner roll
gimbal (2)

Outer roll
gimbal (4)

Azimuth
gimbal (1)

W

V

U

Pitch
gimbal (3)

Y (yaw)

X (pitch)

Flight direction

Prism for optical
alinement

CD-10462-31

Figure VI-70. - Gimbal diagram, AC-16. Launch orientation: inertial platform coordinates, U, V, and W;
vehicle coordinates, X, Y, and Z.

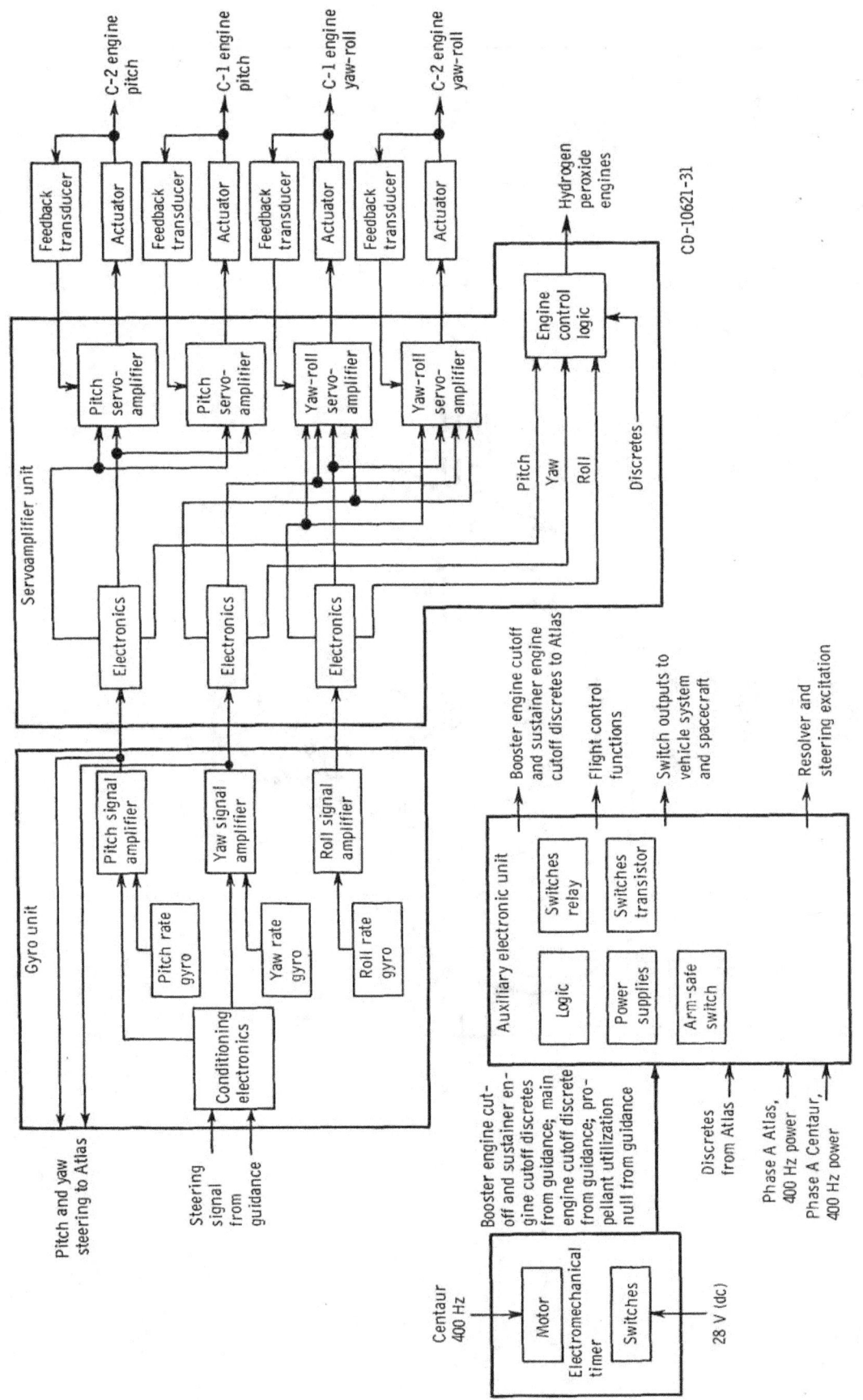

Figure VI-71. – Centaur flight control system, AC-16.

144

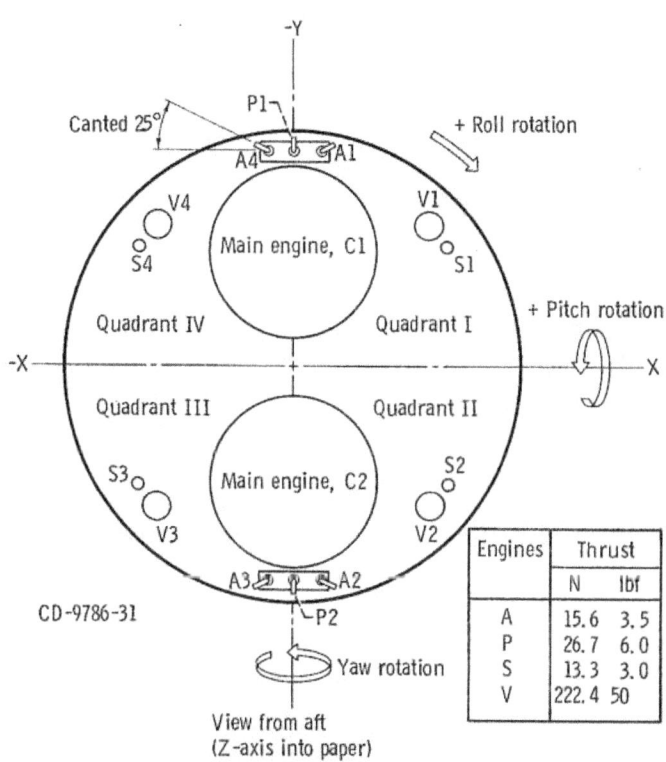

Figure VI-72. - Attitude engines alphanumeric designations and locations, AC-16. Signs of axes are convention for flight control system.

CD-9786-31

Engines	Thrust	
	N	lbf
A	15.6	3.5
P	26.7	6.0
S	13.3	3.0
V	222.4	50

View from aft
(Z-axis into paper)

VII. <u>CONCLUDING REMARKS</u>

The OAO-II, launched in December 1968, was injected into the desired circular Earth orbit at an altitude of 772 kilometers and an inclination of 35^0 to the equator. For this mission the Atlas-Centaur flew a steeper ascent than for any previous Atlas-Centaur flight. All launch vehicle systems performed satisfactorily. The AC-16 flight proved the capability of the Atlas-Centaur to perform with a heavier payload and a longer nose fairing than those flown on previous missions. The OAO spacecraft was larger than the Surveyor spacecraft, requiring a 5.5-meter-longer nose fairing. Payload weight increase over previous missions was about 1000 kilograms.

The objectives of the star-studying observatory OAO-II, carrying 11 telescopes, were to investigate and study the young stars in the ultraviolet spectrum, to map the stars, and to further extend our knowledge of the makeup of the solar system. Many of these stars are not visible from Earth-based observatories because of the distorting effect of the Earth's atmosphere. The success of the OAO-II mission represents a significant breakthrough in the field of observational astronomy.

Lewis Research Center,
National Aeronautics and Space Administration,
Cleveland, Ohio, October 14, 1969,
491-02.

APPENDIX A

SUPPLEMENTAL LAUNCH AND WEIGHTS DATA

by John J. Nieberding

Launch Window and Countdown History

There was no specific launch opportunity, or specified number of days, for the OAO-II mission. Launch could occur on any day. The opening of the launch window on a given launch day was determined by a spacecraft constraint which would not permit spacecraft separation to occur until the spacecraft had been in sunlight for at least 1 minute. This constraint is reflected in the earliest possible lift-off time for the December 7, 1968, launch of 0340 eastern standard time. Launch window durations were chosen by the Goddard Space Flight Center to be 4 hours long.

AC-16 was launched on the first attempt. The countdown started on time at 2055 eastern standard time on December 6, 1968. Planned holds of 60 minutes duration at T - 90 minutes and 10 minutes duration at T - 5 minutes were observed as scheduled. No additional holds or recycles were required, and the countdown proceeded smoothly to vehicle lift-off at 0340:09.165 eastern standard time.

The Atlas and Centaur postflight vehicle weight summaries are presented in tables A-I and A-II.

TABLE A-I. - ATLAS POSTFLIGHT VEHICLE

WEIGHT SUMMARY, AC-16

Item	Weight	
	kg	lb
Booster jettison weight:		
Booster dry weight	2 850	6 283
Booster residuals	472	1 041
Unburned lubrication oil	16	36
Total	3 338	7 360
Sustainer jettison weight:		
Sustainer dry weight	2 700	5 953
Sustainer residuals	388	856
Interstage adapter	482	1 062
Unburned lubrication oil	8	18
Total	3 578	7 889
Flight expendables:		
Main impulse fuel (RP-1)	36 842	81 222
Main impulse oxygen	82 997	182 977
Helium panel purge	2	5
Oxygen boiloff and vent loss	229	504
Lubrication oil	82	182
Total	120 152	264 890
Ground expendables:		
Fuel (RP-1)	640	1 412
Oxygen	1 874	4 132
Lubrication oil	2	5
Exterior ice	24	54
Liquid nitrogen in helium shrouds	113	250
Total	2 653	5 853
Total Atlas tanked weight	129 721	285 992
Minus ground run	2 655	5 853
Total Atlas weight at lift-off	127 066	280 139

TABLE A-II. - CENTAUR POSTFLIGHT VEHICLE

WEIGHT SUMMARY, AC-16

Item	Weight	
	kg	lb
Basic hardware:		
Body	498	1 098
Propulsion group	501	1 105
Guidance group	143	315
Fluid systems group	116	256
Electric group	110	243
Separation group	47	103
Basic flight instrumentation	108	239
Total	1 523	3 359
Mission peculiar hardware, total	933	2 056
Hardware jettisoned in boost phase:		
Insulation panels	526	1 159
Ablated ice	23	50
Total	549	1 209
Hardware jettisoned in Centaur phase:		
Nose fairing	947	2 088
Split fairing and separation equipment	168	370
Total	1 115	2 458
Centaur residuals:		
Liquid hydrogen	86	190
Liquid oxygen	265	584
Gaseous hydrogen	39	85
Gaseous oxygen	71	169
Hydrogen peroxide	88	193
Helium	2	5
Ice	5	12
Total	556	1 238
Centaur expendables:		
Main impulse hydrogen	2 199	4 847
Main impulse oxygen	11 137	24 553
Gas boiloff on ground, hydrogen	8	17
Gas boiloff on ground, oxygen	0	0
In-flight chill, hydrogen	10	23
In-flight chill, oxygen	14	31
Booster phase vent, hydrogen	28	61
Booster phase vent, oxygen	30	66
Sustainer phase vent, hydrogen	14	30
Sustainer phase vent, oxygen	27	60
Engine shutdown loss, hydrogen	3	6
Engine shutdown loss, oxygen	6	13
Hydrogen peroxide	19	42
Helium	1	1
Total	13 496	29 750
Total Centaur weight at lift-off	18 172	40 070
Spacecraft	2 017	4 447
Total Atlas-Centaur-spacecraft weight at lift-off	147 255	324 656

APPENDIX B

CENTAUR ENGINE PERFORMANCE CALCULATION

by Ronald W. Ruedele

The method of calculation used was the Pratt & Whitney C* technique. This technique is an iteration process for determining engine performance parameters. Flight data are used with calibration coefficients obtained from the engine acceptance tests. Calculations are made to determine C*, individual propellant weight flow rates, and subsequently, specific impulse and engine thrust. The procedure is as follows:

(1) Calculate the hydrogen flow rate by using acceptance test calibration data and venturi measurements of pressure and temperature as obtained from telemetry.

(2) Assume a given mixture ratio and calculate corresponding oxidizer flow rate and total propellant flow rate.

(3) Obtain C* ideal from performance curve as a function of mixture ratio.

(4) Correct to C* actual by using characteristic exit velocity efficiency factor obtained from acceptance test results.

(5) Calculate total propellant flow rate, using C* actual;

$$\dot{\omega}_t = \frac{P_o A_t g}{C*}$$

where $\dot{\omega}_t$ is the total engine propellant flow rate, P_o is the measured chamber pressure from telemetry, A_t is the thrust chamber throat area, g is the gravitational constant (32.17 ft/sec/sec), and C* is the characteristic (actual) exhaust velocity.

(6) Determine the mixture ratio by using the calculated total propellant flow rate and the measured hydrogen flow rate.

(7) Compare the calculated mixture ratio with that assumed in step (2).

(8) If two values of mixture ratio do not agree, assume a new value of the mixture ratio and repeat the process until agreement is obtained.

(9) When the correct mixture ratio is determined, obtain the ideal specific impulse from the performance curve as a function of actual mixture ratio.

(10) Correct to actual specific impulse by using specific impulse efficiency factor determined from acceptance test results.

(11) Calculate engine thrust as a product of propellant flow rate and specific impulse.

REFERENCES

1. Gerus, Theodore F.; Housely, John A.; and Kusic, George: Atlas-Centaur-Surveyor Longitudinal Dynamics Test, NASA TM X-1459, 1967.

2. Foushee, B. R.: Liquid Hydrogen and Liquid Oxygen Density Data for Use in Centaur Propellant Loading Analysis. Rep. AE62-0471, General Dynamics Corp., May 1, 1962.

3. Pennington, K., Jr.: Liquid Oxygen Tanking Density for the Atlas-Centaur Vehicles. Rep. BTD65-103, General Dynamics Corp., June 3, 1965.

NEWS

NATIONAL AERONAUTICS AND SPACE ADMINISTRATION
WASHINGTON, D.C. 20546

TELS. WO 2-4155
WO 3-6925

FOR RELEASE: FRIDAY P.M.
November 1, 1968

RELEASE NO: 68-186K

PROJECT: OAO-A2

contents

PRESS

KIT

10/24/68

FOR RELEASE: FRIDAY P.M.
November 1, 1968

RELEASE NO: 68-186

OAO SET FOR LAUNCHING

A star-studying observatory, carrying 11 telescopes designed to investigate the past of the universe so astronomers can better determine its future, is scheduled for launching no earlier than Nov. 12 from Cape Kennedy, Fla.

The National Aeronautics and Space Administration's Orbiting Astronomical Observatory (OAO), weighing 4,400 pounds, is the heaviest and most automated unmanned satellite under development in the United States today.

OAO A2 (OAO-II in orbit) is 10 feet tall and 21 feet wide with its solar arrays unfolded. It contains 328,000 separate parts, nearly three times the amount of the Surveyor spacecraft which successfully landed on the Moon.

A two-stage Atlas-Centaur rocket will place OAO-A2 into a circular orbit at 480 miles above Earth. The observatory's path will be inclined 35 degrees to the Equator. The time for one revolution will be 100 minutes.

OAO-A2 carries two experiments provided by the Smithsonian Astrophysical Observatory and the University of Wisconsin.

Both experiments will observe extremely young hot stars in the ultraviolet-the blue portion of the spectrum not visible to the human eye or Earth-based observatories. Some of these young stars are only tens of thousands of years old. Our Sun is believed to be middle age, about 5 billion years old.

It took 15 years and about 40 sounding rocket flights to obtain approximately three hours of ultraviolet data from some 150 stars. OAO-A2 can collect twice as much ultraviolet information in one day, and from much fainter stars. Over a six-month period, astronomers hope to study more than 50,000 stars.

In addition to young stars, OAO-A2 will observe instellar gas (dust)--the matter from which stars are formed and several planets in our solar system, Mars, Jupiter, Saturn, Neptune and Uranus. Some young stars in the Andromeda nebula and Magellanic clouds may also be studied.

The Andromeda nebula is a galaxy in the constellation Andromeda which is visible in the northern hemisphere. Magellanic clouds are near-galaxies observable only from the southern hemisphere. They look like patchy clouds.

OAO's computer system can store 256 instructions (commands), more than any previous unmanned satellite. Efficient on-board processing of scientific information on OAO will prevent overloading of ground stations that has occurred with previous scientific spacecraft.

Possibly the most significant feature of OAO-A2 capabilities is the pointing precision required to keep its telescopes aimed at a star so astronomers can receive precise scientific data. It has a high-performance coarse pointing accuracy of 1 minute of arc.

This is equivalent to distinguishing between the right or left eye of a person viewed at a distance of 500 feet. Future OAO's will be capable of 0.1 arc second.

Prior to the Space Age, astronomers could see through only two narrow "windows." Optical telescopes--Mt. Palomar for example--were used to see the visible light of stars while radio telescopes listened and measured radio waves emitted from celestial objects.

With OAO satellites, astronomers will have a new
vantage point for seeing in the ultraviolet, infrared, X-rays
and gamma rays, most of the necessary ingredients for knowing
the solar system's make-up today and what course it might
be taking.

Contributions in the visible and ultraviolet, from
ground observatories and sounding rockets respectively,
have greatly increased man's understanding of the solar
system and universe.

Both techniques, however, are inconclusive as to the
origin, evolution and present conditions of the universe.

Sounding rocket ultraviolet measurements cannot go
beyond third-magnitude stars, the typical visible stars.
OAO-A2 experiments can measure much fainter stars,--down to
about the ninth magnitude.

A sounding rocket can observe one star for only a
few minutes, while the Wisconsin experiment on OAO-A2 can
observe a star in some cases for many hours.

Unlike that experiment, which studies one star at a
time, the Smithsonian experiment is much like conducting a
census of people on Earth.

The Smithsonian will survey and produce pictorial maps of more than 700 stars daily. If OAO-A2 lives for about six months, the experiment will photograph some 25 per cent of the sky.

These stellar maps should be as valuable to astronomers and future astronauts as road maps are to travelers on Earth.

Dr. James E. Kupperian, OAO Project Scientist from NASA's Goddard Space Flight Center, Greenbelt, Md., believes a working OAO system would revolutionize observational astronomy.

OAO-A2 is the second in a series of four observatories planned by NASA. OAO-I was launched into an almost perfect orbit April 8, 1966 but failed due to a malfunction in the power supply system and probable high voltage arcing in the star tracker. Several modifications to the OAO system have been incorportated into the OAO-A2 spacecraft as a result of the first flight.

OAO-B and C, scheduled for flights in late 1969 and late 1970, will carry new experiments and, because of the experiment requirements, will have pointing systems accurate within 0.1 arc second.

OAO-B will carry the Goddard 38-inch aperture telescope. OAO-C will have the Princeton University 32-inch aperture reflecting telescope.

The Orbiting Astronomical Observatory program is directed by NASA's Office of Space Science and Applications. Project management is under Goddard Space Flight Center.

Launch vehicle management is at the Lewis Research Center, Cleveland, and launch operations are under the direction of the Kennedy Space Center, Fla.

Grumman Aircraft Engineering Corp., Bethpage, N. Y., is prime contractor for the OAO spacecraft. Electro-Mechanical Research, Inc., Princeton, N. J., built the Smithsonian experiment and Cook Electric Co., Morton Grove, Ill., made the Wisconsin experiment.

Prime contractor for the Atlas-Centaur launch vehicle is General Dynamics-Convair, San Diego. More than 1,000 major subcontractors and vendors participated in the OAO-A2 effort.

OAO-A2 FACTS

Spacecraft

Weight:
4,400 pounds (1,000 pounds of scientific experiments)

Shape:
Octagonal cylinder, 7 feet wide, 10 feet long, wing span of 21 feet with solar paddles deployed, resembles a bat in flight

Stabilization-Control:
Six star trackers with a 1-minute arc accuracy; two are required at one time

Experiments

Smithsonian Astrophysical Observatory:
Survey the young, hot ultraviolet stars with four telescopes. A major contribution from this experiment could be detecting new stellar objects;

University of Wisconsin:
Will observe about 15 stars daily, with emphasis on studying one star at a time for extended periods (up to several hours).

Launch Information

Vehicle:
Two stage Atlas-Centaur

Complex:
36B, Cape Kennedy

Azimuth:
60 degrees True

Window:
3:15 to 5:15 a.m. EST (approx.)

Date:
November 12, 1968

Orbital Elements

Apogee-Perigee:
Circular, 480 statute miles

Period:	100 minutes
Inclination:	35 degrees
OAO-A2 Ground Stations	Rosman, N.C.; Quito, Ecuador; Santiago, Chile; Orroral, Australia; Tananarive, Malagasy
Spacecraft Management	Goddard Space Flight Center Greenbelt, Md.
Launch Vehicle Management	Lewis Research Center, Cleveland
Launch Operations	Kennedy Space Center, Fla.
Prime Contractors	
Spacecraft:	Grumman Aircraft Engineering Corp., Bethpage, N.Y.
Launch Vehicle:	General Dynamics-Convair, San Diego

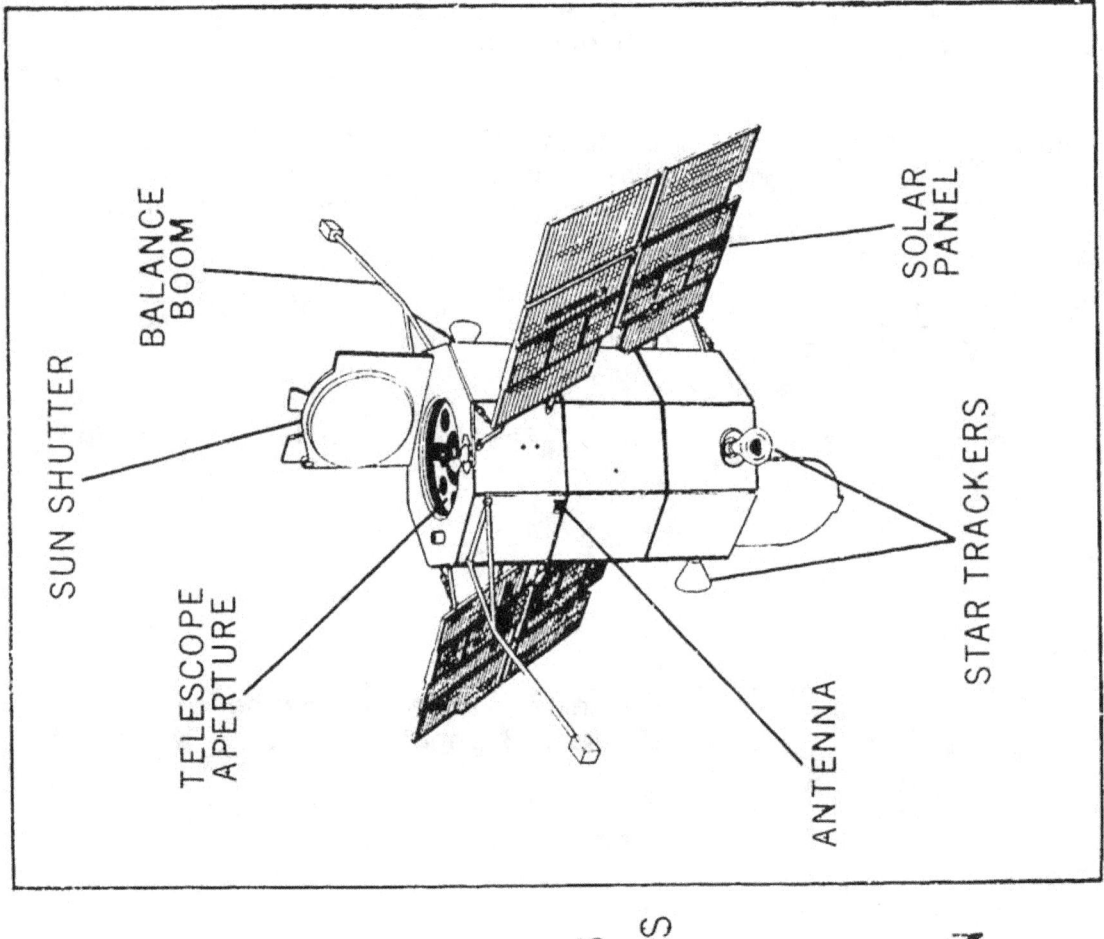

ORBITING
ASTRONOMICAL
OBSERVATORY-A2

GROSS WT. — 4,400 LBS.

INSTRUMENT WEIGHT — 1,000 LBS.

INSTRUMENT PAYLOAD — 11 TELESCOPES

STABILIZATION — ACTIVE 3-AXIS (6 STAR TRACKERS)

POINTING ACCURACY — 1 ARC MIN.

ORBIT — CIRCULAR, 480 S. MI. INCLINATION 35°

LAUNCH VEHICLE — ATLAS/CENTAUR

SUN SHUTTER

BALANCE BOOM

SOLAR PANEL

TELESCOPE APERTURE

STAR TRACKERS

ANTENNA

NASA G-69-2065

164

SPACECRAFT COMPLEXITY

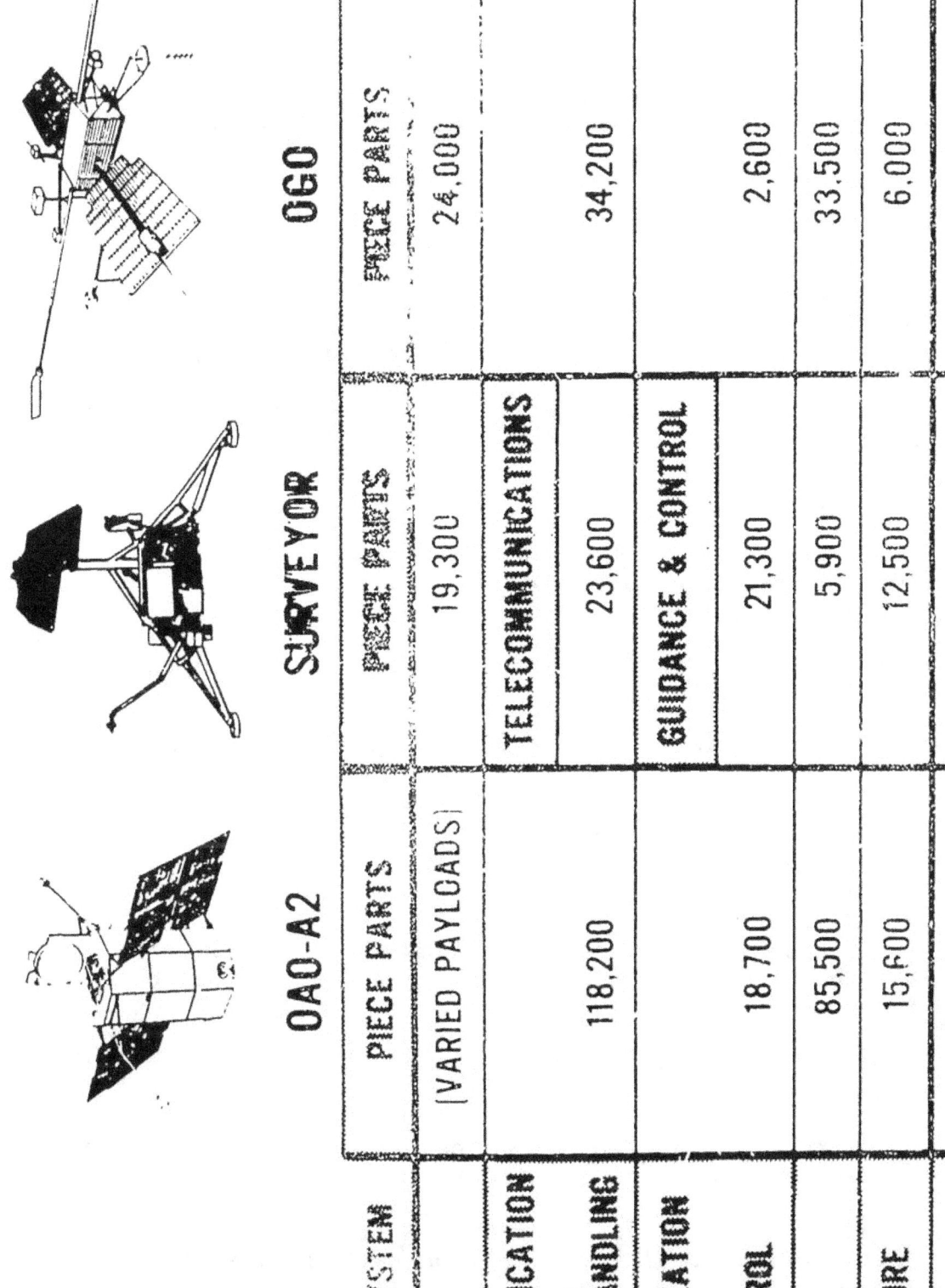

SUB-SYSTEM	OAO-A2 PIECE PARTS	SURVEYOR PIECE PARTS	OGO PIECE PARTS
SCIENCE	(VARIED PAYLOADS)	19,300	24,000
COMMUNICATION & DATA HANDLING	118,200	TELECOMMUNICATIONS 23,600	34,200
STABILIZATION & CONTROL	18,700	GUIDANCE & CONTROL 21,300	2,600
POWER	85,500	5,900	33,500
STRUCTURE	15,600	12,500	6,000
TOTAL PARTS	328,000	82,600	100,300

OAO SCIENTIFIC OBJECTIVES

Since the invention of the telescope in the 17th Century, astronomers have collected substantial amounts of new knowledge from stars.

More recently, astronomers have received stellar information from balloons, aircraft and sounding rockets.

OAO spacecraft will take the astronomers' instruments above the distorting effect of the Earth's atmosphere for detailed observations of the entire electromagnetic spectrum.

Some of the areas which have long interested astronomers, and which OAO-A2 will investigate in great detail for the first time in the ultraviolet:

- Study giant stars with masses more than 100 times as large as our Sun. These stars should blow up with violent explosions, or Supernovae. Why don't they blow up?

- Study stellar temperature of young stars to learn more about their ages.

- Study helium content of stars to determine nuclear processing which has gone on in recent years.

- Study stellar chromosphere (outer edges of stars). Are the chromospheres more active in the giant stars or the dwarfs?

- Study the Red Giant stars, which in their late stages, have burned most of their hydrogen, causing their cores to contract.

- Study various types of stars to see how their chemical composition differs.

- Study the origin, evolution and structure of the more massive, hot stars 1 million years old.

- Discover new classes of objects which might be brighter in the ultraviolet than current theories predict.

- Study interstellar matter, from which stars are born, to determine the amount of absorption between stars and Earth which would provide clues on the amount of dust in space.

- Study stellar hurricanes, which are blowing from stars millions to billions times more intensely than solar wind.

- Study young stars in spiral arms of other galaxies.

- Make observations which will assist in the difficult problem of determining the internal configurations of stars.

- Study the color and distribution of unusual stars.

OAO SPACECRAFT

OAO-A2 is an eight-sided spacecraft 10 feet long and seven feet wide. Eight sets of panels, consisting of two arrays of four panels each covered with more than 109,000 solar cells, give the observatory an over-all width of 21 feet.

The internal structure of the main body includes a four-foot-diameter central tube where the experiments are mounted, surrounded by vertical trusses and horizontal shelves. The bays formed by the trusses and shelves provide space for mounting spacecraft electronics and data handling systems.

OAO-A2's main body is constructed of riveted or spot-welded aluminum alloy. Extensive use has been made of aluminum honeycomb in locations where high rigidity is needed.

A thin non-structural covering of specially fabricated aluminum coated with Alzak (an aluminum oxide) covers the main body except for the experiment opening. The treated-aluminum-covering is a vital part of the passive thermal control system.

Attitude and Control System

Success of the OAO-A2 mission depends on the ability of the 4,400-pound observatory to point its astronomical instruments at pre-selected objects in space. The OAO attitude control system is one of the most advanced ever developed.

Star Trackers

Six gimbaled star trackers are the heart of the system.

After the observatory separates from the Centaur stage, the spacecraft will go into an automatic "sun bathing" mode stabilizing on the Sun.

The first star tracker will be turned on in orbit No. 25. The remaining five trackers will be turned on during orbits 27 and 28.

The equipment used to sense OAO motions consists of rate gyros to measure initial tumbling rates (after spacecraft separation), solar sensors to establish the direction of the Sun, the six star trackers and one bore-sight star tracker.

The star trackers are designed to acquire selected guide stars, track them continuously and at the same time measure their direction with respect to the spacecraft axes.

The following stars and constellations have been selected as guide stars for OAO-A2's star trackers:

Sirius	Adara	Kaus Australis
Carina	Aried	Orion
Vega	Bellatrix	Alhena
Rigel	Castor	Sagittarius
Achernar	Alioth	Grus
Centaurus	Nath	Sirrah
Capella	Mirzam	Arcturus
Spica	Crux	Fomalhaut
Procyon	Alkaid	Aldebaran
Altair	Vela	
Regulus	Pavo	

To initiate control maneuvers, the spacecraft houses a nitrogen gas jet system -- used primarily for initial stabilization -- a coarse momentum wheel system for star tracker control.

The key to the OAO control system is the star tracker system. It must point the observatory to an accuracy of one minute of arc and maintain this pointing direction within 15 arc-seconds for 50 minutes.

This accuracy is needed to assure that desired "target" stars fall within the field of view of the experiments.

Each star tracker is a small 3.5-inch reflecting telescop mounted in two degree-of-freedom mechanical gimbals.

The incoming target star image is split into two light beams to provide error signals about the two gimbal axes. The beams are modulated by a system of vibrating reeds, detected by a photomultiplier and electrically separated into error signals.

The resulting error signals are then used to drive direct current torquer motors in the gimbal axes. Gimbal angles are measured by variable capacitance transducers with a resolution of about five arc-seconds.

Two trackers are sufficient to provide pointing information under normal operating conditions.

However, six trackers are used to allow for occultation of guide stars by the Earth, to maintain proper reference when the spacecraft shifts guide stars and for redundancy to improve the lifetime of the observatory.

OAO-A2 SCIENTIFIC EXPERIMENTS

The Smithsonian Astrophysical Observatory and the University of Wisconsin provided the two experiments for the OAO-A2 mission.

Both experiments will measure radiation in the ultra-violet portion of the electro-magnetic spectrum.

While both experiments involve the ultraviolet, their assignments differ. The Smithsonian will map, or survey, 700 stars daily providing astronomers with their first detailed ultraviolet stellar map.

The Wisconsin Experiment Package will study, in great detail, one star at a time to define better the chemical composition, pressure and density of stellar objects. This information could result in revision of present theories of stellar origin and revolution.

The two experiments can complement each other. If Smithsonian should discover a new star or object, Wisconsin could then "zero in" for a close-up look.

If OAO lives for six months, Smithsonian will have charted as many as 50,000 stars and Wisconsin more than 1,000.

Celescope   Experimenter: Dr. Fred Whipple
                                   Smithsonian Astrophysical
                                   Observatory, Cambridge, Mass.

Celescope (celestial telescope) will measure the brightness of young stars in four spectral bands between 1,000 and 3,000 angstroms (an angstrom is about 254-millionths of an inch).

The entire Smithsonian experiment weighs about 500 pounds.

Celescope will measure with four large-aperture television cameras using broad-band television photometers, significant numbers of O, B, A and F stars in the longer wavelengths and O, B and A stars in the shorter wavelengths.

A special type of television tube, called the "uvicon," sensitive only to ultraviolet lights, was developed for Celescope.

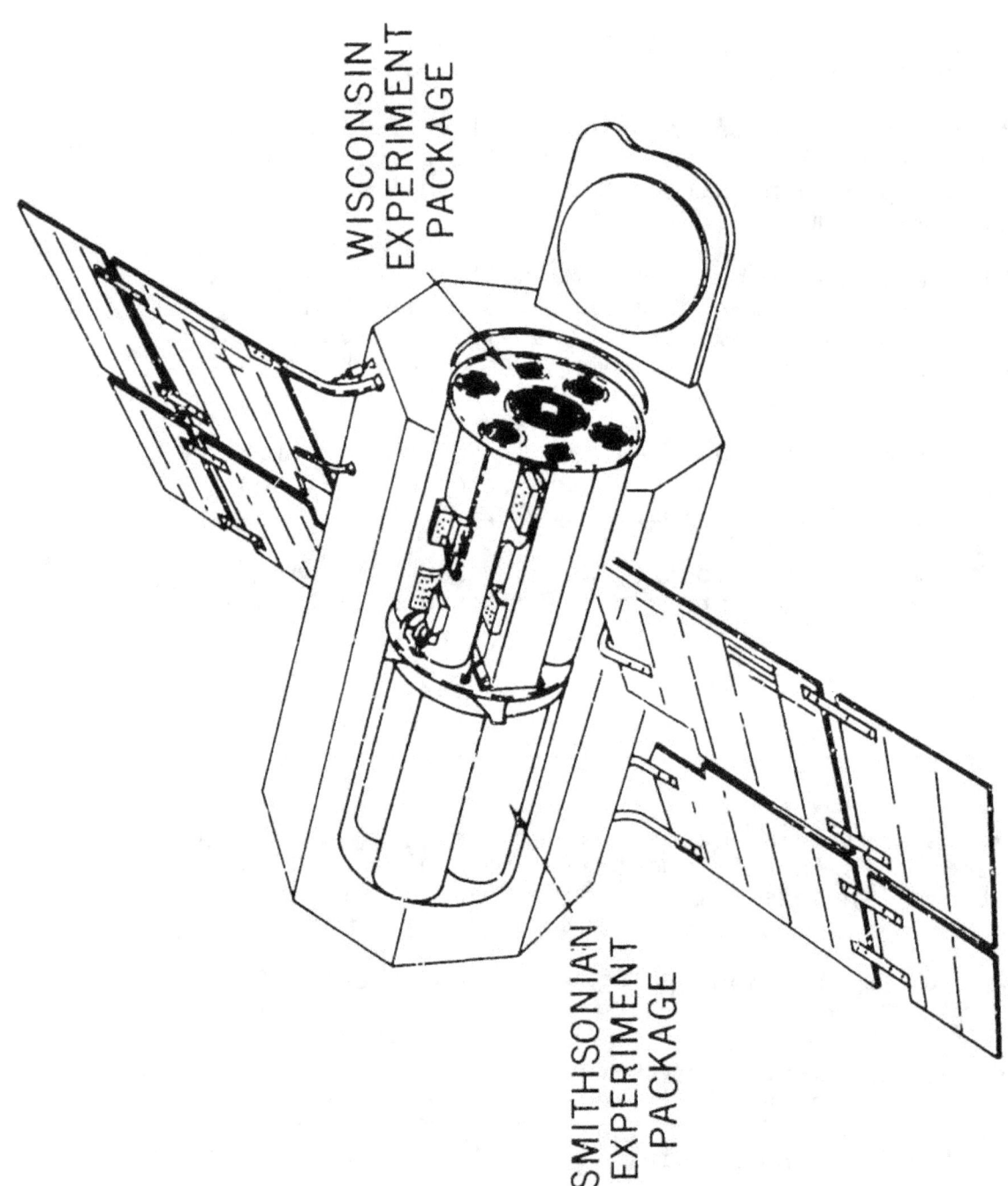

WISCONSIN EXPERIMENT PACKAGE

SMITHSONIAN EXPERIMENT PACKAGE

PHANTOM VIEW OF OAO-A2 EXPERIMENTS

NASA G-69- 1713

In six months Celescope will have catalogued more than 25 per cent of the sky (about 50,000 stars) down to magnitude 9, or about 12 times fainter than what the human eye can see.

Optical data collected by Celescope will be converted into numerical data by instrumentation aboard OAO-A2 and transmitted back to Earth.

On the ground the data will be analyzed, and with the aid of computers, converted into photographs, new maps and catalogs of the celestial sphere, as it looks in the ultra-violet.

If possible, exposure of the television cameras will be increased to permit ultraviolet observation of other objects, such as bright nebulosites (large class of celestial struc-tures composed of matter in a gaseous or finely divided state), radio sources, planets (including Earth), and perhaps the newly discovered quasars.

Quasars, until 1963, were thought to be faint stars in the Milky Way. They are considered among the most distant celestial objects man has observed. The farthest quasars are now believed to be some 4,000 million light years from the Milky Way.

More than 30 quasars have been identified and several new ones are expected to be discovered.

Celescope will be turned on about the eighth day in orbit. The first three days of stellar pictures will be taken in the Constellation Draco, an area of the sky which is not very crowded with stars.

After Wisconsin has the spacecraft for one week, Celescope will take pictures in a very crowded region in the southern Constellation Scorpius (in the Milky Way).

When the OAO-A2 has reached the plane of the Milky Way, about one month after launching, Celescope will begin photo-graphing 700 stars daily.

Smithsonian plans to publish the first ultraviolet catalogue about six months after launching.

The four, identical high resolution telescopes, with 12.5-inch diameters, four filters and four image-forming uvicon tubes will image sections of the sky about 2.8 degrees in diameter in four spectral ranges.

The Uvicon signals will be quantized into 128 levels, providing an accuracy of about 0.1 magnitude. A complete image requires the transmission of about a half-million bits of information. Between one and four photos are planned per orbit, depending on the length of time OAO is over ground stations.

Normal exposure times will be about one minute.

Wisconsin Experiment Package   Experimenter: Dr. Arthur D. Code
University of Wisconsin,
Madison, Wisc.

The Wisconsin experiment will be the first large scale effort to obtain detailed ultraviolet photometric measurements on stars and nebulae.

Comprised of seven separate photometric systems, it will study selected stars, especially the hot young ones, in the 1,000-3,300-angstrom region for extended periods.

An average of one star per orbit will be studied. If OAO-A2 operates for six months, Wisconsin could study more than 1,000 stars.

In addition to stellar study, Wisconsin hopes to obtain the first ultraviolet information on instellar gas (dust) in the Orion Nebula (Orion Sword). It will measure line emissions which light up in the ultraviolet like a neon sign.

Multiple observations will be made on selected targets over a period of time to check observational consistency and to verify that the energy distribution is not intrinsically variable.

The Wisconsin experiment will be turned on for the first time about five days after launching. Over a two-day checkout period, it will take photometric measurements from six stars including the well known Canopus which has been used as a navigational star for several of NASA's deep space missions.

The Wisconsin experiment might obtain information, within the first few days of operation, on Olber's paradox which concerns cosmology (theory of the creation of the universe).

Olber's paradox states that if the universe were infinite, an individual on Earth could look in any direction, and eventually this line of sight would intersect a star. If this were the case, the entire sky would be as bright as the Sun, and the temperature on Earth would be some 6,000 degrees.

Why, then, is the sky dark at night? Astronomers say this isn't easy to understand unless the assumption is made that the universe is finite.

The experiment, weighing 450 pounds, consists of three separate types of instruments -- four stellar photometers, a nebular photometer and two scanning spectrographs.

The stellar multicolor photometer system, intended primarily for measurement of stars, consists of four eight-inch telescopes, each sending information to a separate three-color filter photometer.

The multicolor filter photometer system, designed primarily to study nebulae, consists of a 16-inch telescope.

The scanning spectrometer system employs two objective grating spectrometers with an aperture of about six by eight times.

In general, the experiment works as follows: The stellar photometer-telescopes and associated mechanisms measure the intensity of incoming ultraviolet light and convert these measurements into electrical signals. By using a rotating filter wheel, measurements at different wavelengths are obtained. The nebular photometer performs similarly.

The spectrometer spreads the star light into a "rainbow" allowing the independent measurement of various wavelengths (colors) without the need for filters.

The experiment optics are protected by a sunshade located at the top of the spacecraft.

During the launch phase the sunshade will be closed over the experiment tube. After orbit is attained, the sunshade will be opened to permit experiment operation.

If the OAO-A2 control system is inadvertently pointed toward the Sun, the shade will close automatically to keep out potentially damaging solar rays.

OAO-I

OAO-I was successfully launched from Cape Kennedy into an almost perfect orbit about 500 miles high Apr. 8, 1966. After several days, the mission was declared a failure due to an electrical malfunction.

Causes of the OAO-I major problems have been attributed to a failure in the power supply system and probable high voltage arcing in the star trackers.

In its 20-orbit lifetime OAO-I successfully performed spacecraft operations in several complex areas, which are required in an OAO system.

The spacecraft performed its sequence of stabilization and control operation, consisting of rate stabilization after separation; coarse and fine solar pointing; roll-search, and coarse pointing wherein the star trackers provide a celestial reference system for spacecraft stabilization.

This stabilization was achieved on several orbits.

The spacecraft command system received, verified, stored, and executed commands accurately.

The first sign of a potential problem occurred about eight minutes after the observatory separated from the Agena rocket. It is believed than an electrical transient occurred, after initial star tracker turn-on, which affected the status of some of the spacecraft subsystems.

A second star tracker experienced an unscheduled turn-on and the stabilization and control mode of roll-search was terminated.

Subsequent to this electrical transient, data indicated that all equipment resumed normal operation except for a loss of some channels of spacecraft status data.

The two major problems that seriously affected the mission operations were the inability to program spacecraft operations reliably because of spurious clock resetting, and the high battery temperature which shortened the battery lifetime.

Battery power was lost after the 20th orbit and no further spacecraft data were received. The experiments were never activated.

-more-

Several required modifications have been made to OAO-A2 to insure mission success.

The battery charging system has been redesigned to improve system reliability and redundancy, to simplify logic circuitry, to parallel battery outputs and to provide ground command control of the battery charge control circuits and batteries.

Modifications have been made to the star tracker system to prevent high voltage induced arcing and corona discharge.

In addition, the intensive investigations by NASA of the OAO-I program development, test and operations history have resulted in design and program modifications directed toward increasing the reliability and operating efficiency of the OAO system in order to insure the success of the second OAO mission. Such changes were within the schedule of the pacing modifications established by the two design changes required.

ATLAS-CENTAUR LAUNCH VEHICLE

The Orbiting Astronomical Observatory, weighing 4,436 pounds, is the heaviest payload the Atlas-Centaur (AC-16) has been called on to launch. On the OAO mission, Centaur's task is to place the spacecraft into a circular orbit 480 statute miles above the Earth. After separation, Centaur must be moved into another orbit to avoid confusing the instruments on the spacecraft.

The first seven operational missions for Centaur were launches of Surveyor spacecraft to the Moon. This very successful series of launches has proved the great value high energy upper stages can have in the space program. Centaur, which was developed under the direction of NASA's Lewis Research Center, Cleveland, was the first U.S. vehicle to use the liquid hydrogen-liquid oxygen propellant combination.

AC-16 consists of an Atlas SLV-3C booster combined with a Centaur second stage. The two stages are 10 feet in diameter and are connected with an interstage adapter. Both Atlas and Centaur stages rely on internal pressurization for structural integrity.

The Atlas booster develops 395,000 pounds of thrust at liftoff using two 168,000 thrust booster engines, one 58,000 thrust sustainer engine and two vernier engines developing 670 pounds thrust each.

The Centaur second stage including the nose fairing is 65 feet long. It is powered with two improved RL-10 engines, designated RL-10, A-3-3. The RL-10 was the first operational hydrogen-fueled engine developed for the space program.

Centaur carries insulation panels and a nose fairing which are jettisoned after the vehicle leaves the Earth's atmosphere. The insulation panels, weighing about 1,200 pounds, surround the second stage propellant tanks to prevent the heat of air friction from causing excessive boil-off of liquid hydrogen during flight through the atmosphere. The nose fairing protects the payload from the same heat environment.

Launch Vehicle Characteristics

Liftoff weight including spacecraft:*	319,918 pounds
Liftoff height:	135 feet 4 inches
Launch Complex:	36 B
Launch Azimuth:	60 degrees

	SLV-3C Booster	Centaur Stage
Weight:*	279,906 lbs.	40,012 lbs.
Height:	70 feet	65 feet, 4 inches (with payload fairing)
Thrust:	395,000 lbs. (sea level)	30,000 lbs. (vacuum)
Propellants:	Liquid oxygen and RP-1	Liquid hydrogen and liquid oxygen
Propulsion:	MA-5 system (2-168,000 lb. thrust engines, 1-58,000 lb. sustainer engine and 2-670 lb. thrust vernier engines.)	Two 15,000 pound thrust RL-10 engines. 14 small hydrogen peroxide thrusters.
Velocity:	5335 mph at BECO 6452 mph at SECO	15,745 mph at s/c separation
Guidance:	Pre-programmed auto-pilot through BECO. Switch to Centaur inertial guidance for sustainer phase.	Inertial guidance

*Measured at two inches of rise.

Several changes had to be made to prepare Centaur for the OAO launch. The major changes were redistribution of the weight of the payload on the forward end of the Centaur tank and a new longer nose fairing to accommodate the large spacecraft.

The new nose fairing is 32 feet long, approximately 10 feet longer than the fairing used for Surveyor. It is of standard fiberglass construction, using 16 explosive bolts and a spring separation system to jettison after its flight through the atmosphere. Nose fairing jettison tests were conducted in the Space Power Chamber at Lewis Research Center to help qualify the new shroud for flight.

The increased length of the launch vehicle and nose fairing and the change in mass distribution made it necessary to move the Atlas displacement gyro package from its usual location in a pod to the nose of the Atlas booster and moving the Atlas rate gyro five feet forward on the Atlas.

FLIGHT SEQUENCE

Atlas Phase

After liftoff, AC-16 will rise vertically for about 15 seconds before beginning its pitch program. Beginning at two seconds after liftoff and continuing until T+20 seconds the vehicle will roll to the desired flight azimuth of 60 degrees. The roll program on AC-16 takes five seconds longer than usual because of the required inclination of the final orbit.

After 153 seconds of flight, the booster engines are shut down (BECO) and jettisoned. The Centaur guidance system then takes over flight control. The Atlas sustainer engine continues to propel the AC-16 vehicle to an altitude of 184 miles. Prior to sustainer engine shut down, the second stage insulation panels are jettisoned.

The Atlas and Centaur stages are then separated by an explosive shaped charge that slices through the interstage adapter. Retro-rockets mounted on the Atlas slow the spent Atlas stage.

Centaur Phase

At four minutes five seconds into the flight, the Centaur's two RL-10 engines are ignited for a planned seven minute eight second burn. This will place Centaur and the spacecraft into a near circular orbit at an altitude of approximately 480 miles.

Twelve seconds after main engine start, the nose fairing around the spacecraft is separated. At main engine start plus 15 seconds, Centaur initiates a right yaw maneuver to attain the final orbital inclination of 35 degrees. The original launch azimuth was 60 degrees to avoid the Bermuda area during reentry of the Atlas sustainer engine and tank and the nose fairing.

Separation

Separation of the OAO spacecraft takes place by firing explosive bolts on a V-shaped metal band holding the spacecraft to the adapter. Compressed springs then push the spacecraft away from the launch vehicle at a rate of about 3.2 feet per second.

OAO-A2 MISSION PHASES

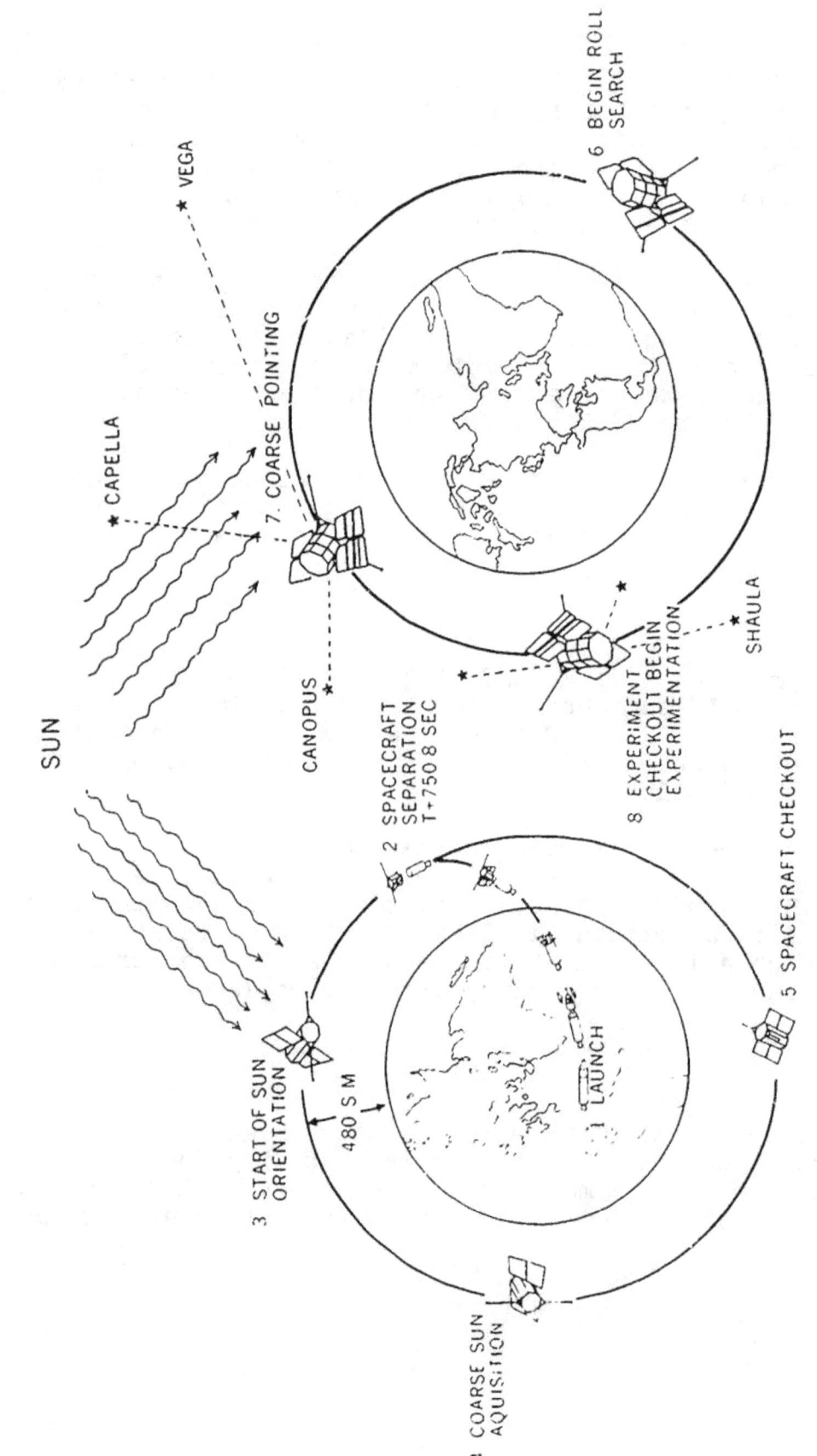

SUN

CAPELLA ★

★ VEGA

7. COARSE POINTING

CANOPUS ★

★ SHAULA

2 SPACECRAFT
SEPARATION
T+750 8 SEC

8 EXPERIMENT
CHECKOUT BEGIN
EXPERIMENTATION

6 BEGIN ROLL
SEARCH

3 START OF SUN
ORIENTATION

480 S M

5 SPACECRAFT CHECKOUT

1 LAUNCH

4 COARSE SUN
AQUISITION

ROLL SEARCH TO COARSE POINTING

ASCENT PROFILE

NASA G-69-2064

182

Retro Maneuver

Five minutes after spacecraft separation, the Centaur stage attitude control thrusters are used to reorient the vehicle. The 50-pound thrust vernier engines are then fired to settle the propellants. The remaining liquid oxygen and liquid hydrogen are vented overboard to provide enough thrust to place the Centaur stage in a slightly different orbit from the spacecraft. The Centaur stage may be visible to the spacecraft for 8-12 days. It will then be out of sight of the spacecraft for 92-138 days and then back in view a period twice as long as the original period during which it was visible to the spacecraft. This cycle will then be repeated at the same intervals.

The final orbital period of the Centaur stage will be only a few seconds longer than that of the spacecraft. The spacecraft period will be approximately 100.29 minutes.

The final Centaur orbit will have an apogee of 504 miles and perigee of 456 miles.

Launch Window

The OAO-A2 launch window opens at approximately 3:17 a.m. EST, Tuesday, Nov. 12 and closes about 2 hours later. In case of delays, the window opens 30 seconds later each day. The window is calculated so that the Sun will be low on the horizon as viewed from the spacecraft at separation.

Atlas-Centaur Flight Sequence

EVENT	NOMINAL TIME MINUTES, SECONDS	ALTITUDE STATUTE MILES	SURFACE RANGE, STATUTE MILES	VELOCITY MPH
Liftoff	0	0	0	0
Booster Engine Cutoff	2'33"	52.3	35.3	5335
Booster Jettison	2'36"	55.9	38.1	5376
Jettison Insulation Panels	3'18"	104.4	80.3	5819
Sustainer Engine Cutoff	3'55"	148.2	123.3	6452
Atlas Separation	3'57"	150.6	125.8	6426
Centaur Engine Start	4'06"	161.9	137.9	6278
Jettison Nose Fairing	4'18"	176	153	6284
Centaur Engine Cutoff	11'25"	483.9	1102	15,740
Spacecraft Separation	12'15"	484	1296	15,745
Start Centaur Reorientation	17'22"	484.1	2486.2	15,743
Start Centaur Retro-thrust	18'57"	483.9	2854.3	15,744

SPACECRAFT ORBITAL OPERATIONS

Orbital operation of OAO-A2 is divided into four phases. They are "survival" (0-1 days), observatory checkout (2-10 days), initial experiment operation (11-24 days) and extended operations (25 days - observatory lifetime).

Survival

In this phase all procedures are directed toward assuring survival of the spacecraft.

The command memory is programmed automatically to carry the spacecraft to the "sunbathing" attitude with a minimum expenditure of gas, and without aid of ground stations.

Operations personnel will analyze data to determine that spacecraft stabilized, battery charging is safe, thermal conditions are within predictions and the solar array output (power system) is normal.

Observatory Checkout

This checkout, a comprehensive shakedown of the spacecraft, experiments and ground stations will begin on the second day and end on the 11th day (orbits 16 through 139).

Initial turn-on of all spacecraft subsystems, including experiments, will occur while the observatory is still being automatically controlled in the sunbathing mode.

During spacecraft checkout, extreme caution will be taken prior to turning on high voltage subsystems. This permits proper outgasing and reduces the likelihood of "arcing" or "corona."

High voltage systems will be turned on initially only when the spacecraft is under Rate And Position Sensor control. With this automatic control, the spacecraft's attitude remains fixed for extended periods without having to use the star trackers.

The first star tracker high voltage will be turned on in orbit 25. The other five trackers will be turned on in orbits 27 and 28.

185

OAO-A2 RATE AND POSITION SENSING SYSTEM (RAPS)

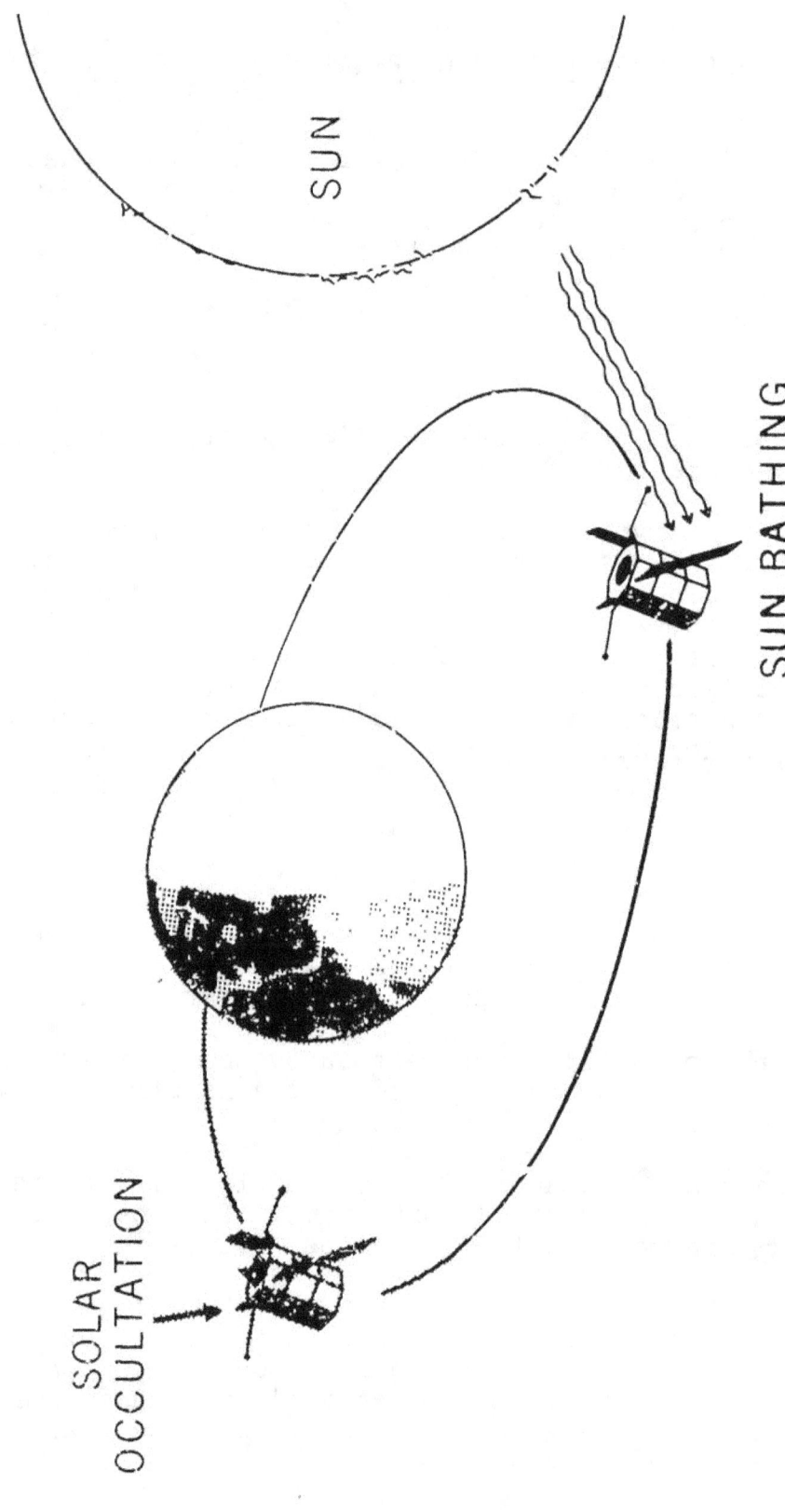

SUN

SOLAR OCCULTATION

SUN BATHING

NASA G-69-1712

Additional high voltage turn-ons will be the Wisconsin and Smithsonian experiments on orbits 56 and 85 respectively.

Experiment checkout is scheduled for six days (orbits 56-139). Wisconsin will do observations and photometry of six stars in a two-day checkout period (orbits 56-84). Smithsonian experiment will be exercised and analyzed from orbits 85 to 139.

Initial Experiment Operations

This phase will continue from 11th through 24th day. Observatory pointing will be specified by the experimenters during this period.

Extended Operations

Objective of this phase, which begins about 25 days after launch, is to obtain as much astronomical information as possible and to continually improve the operational efficiency of the OAO space-ground systems.

Officials at Goddard will attempt to improve the efficiency of experiment scheduling. ground station operations, and data analysis so a maximum of scientific information can be obtained without endangering spacecraft performance.

-more-

ASTRONOMY BACKGROUND

Birth

The theory of most astronomers is that the first stage in stellar formation is contraction, due to gravitational force, of a very large volume of gas, mainly hydrogen and dust.

During contraction, turbulence may cause the large mass to break up into smaller parts leading eventually to a group of stars or clusters.

Because of the gravitational contraction, the surface temperature of the proto-star (as it is now called) has greatly increased. Its brightness has increased less rapidly because the radius has steadily decreased.

The temperature during the early stages would still be too low for radiation to be detectable in the visible part of the spectrum. However, the proto-star would emit energy in infrared.

Growing Up

In about 15 years the contracting proto-star becomes hot enough (2,200 degrees F.) to become visible. The more massive the proto-star the more rapidly it contracts and becomes visible.

Shortly after the young star becomes visible, depending on its mass, it stabilizes and is called a main sequence star.

The young, bright stars emit most of their energy in ultraviolet. Their chemical composition is about 80 per cent hydrogen, 20 per cent helium, and one part in 1,000 contaminants (carbon, nitrogen and oxygen).

These contaminants provide electrons to the stellar atmosphere, causing the star to heat up because the radiation cannot escape as easily.

At least 90 per cent of the known stars are main sequence stars. They will remain in this category until about 20 per cent of their hydrogen is turned up.

-more-

When this happens the hot core of the star, where the thermonuclear reactions have been occurring, is mostly helium. The energy-producing processes in the star's interior have changed and it begins to move away from the main sequence category.

Stars with masses 1.1 to 1.4 times greater than the Sun are generally understood. However, the destiny of the large, massive stars is rather uncertain.

Most of the stars not in the main sequence are red giants (super-giants) and white dwarfs.

Red Giants

These giant stars, ranging in surface temperature from about 5,400 to 7,500 degrees F., are either orange or red in color. They are from a hundred to several thousand times more luminous than main sequence stars of the same spectral class.

The difference in the ratio of brightness increases with decreasing temperature. This means that the brighter stars must have larger radii than those on the main sequence with identical temperatures.

The red giant Capella (Auriga constellation) is as hot as the Sun and its brightness is about the same. However, because the star's diameter is some 12 times greater than the Sun, it emits about 150 times as much light.

The red super-giants, such as Betelgeuse (pronounced beatle juice), have even larger radii than the red giants with the same surface temperatures.

Betelgeuse, in the constellation Orion, has a radius of 180 million miles. It would stretch from the center of the Sun to the Earth, and some 90 million miles beyond.

Astronomers have generally concluded that the temperatures of red giant, and super-giant, stars are several hundred degrees less than main sequence stars of the same spectral class.

White Dwarfs

The white dwarfs are very small and heavy, compared to other stars. They make up about 10 per cent of all the stars in the galaxy. They are very faint, about 10-15 magnitude.

Astronomers need more background on their evolution, so new light can be shed on their past from ultraviolet experiments aboard OAO-A2.

Some scientists believe that white dwarfs are the final evolutionary phase of stars, and that they are the product of an intermediate stage called nova, the Greek word for "new".

White dwarfs are faint blue, white or yellow. Those in the same spectral class as the Sun have a radius like the Earth but a mass about the same as the Sun.

They are very dense and made up of degenerate gas. Although the radii of Earth and the average white dwarf are about the same, a volume of one cubic inch for white dwarfs weighs 30-40 tons.

Several hundred white dwarfs have been noted, but many more will be discovered.

Although they are some 2,500,000 light years from Earth, it is possible that OAO-A2 might discover novae in the galaxy next to ours, Andromeda.

A possible explanation of novae event (stellar blow-up less intense than supernovae) is that during the star's evolution helium and hydrogen are exhausted and the rate of heat being generated from the star decreases. Gas degeneracy then sets in and extends almost to the exterior of the star.

Novae never have been observed in the ultraviolet. Because about 20-30 novae occur in our galaxy annually, OAO-A2 has a 10 per cent chance of making a detailed observation.

Supernovae

The evolution of a star from proto-star, to main sequence, to white dwarf pertains only to those objects about 1.1 to 1.4 the mass of our Sun.

Stars smaller than the Sun may behave the same way, but not the large massive ones several times as large as the Sun.

One theory is that massive stars completely tear themselves up, or disintegrate, into supernovae due to the internal temperature increasing so rapidly that thermonuclear reactions involving helium occur.

The intensity of a supernova is several billion times greater than a hydrogen bomb explosion on Earth.

A supernova is a massive star which becomes extremely bright in a relatively short time by a factor of several hundred millions. The bright phase -- brighter than any of the planets in our solar system except the Sun and Moon -- may last a year or two.

Some astronomers believe that the bright star mentioned in Biblical history by the three wise men some 2,000 years ago was a supernova.

About 50 supernovae have been officially recorded -- three in our galaxy, in the years 1054, 1572 and 1604. One supernova occurs per galaxy in every 50-300 years.

No supernovae have occurred in our galaxy since the invention of the telescope in the 17th Century.

The Chinese and Japanese recorded one of the better known supernovae in the Milky Way in 1054. The remains of the event is the Crab nebula in the constellation Taurus.

Draper's Classification

Most stars can be divided into seven main groups by the letters O, B, A, F, G, K and M. A few stars have spectra that fall into special categories, R-N and S.

This sequence is called the Draper (or Harvard) classification, after the American astronomer Henry Draper (1837-1882), who made important contributions to stellar spectroscopy.

Each class is further sub-divided to permit finer gradations by adding a numeral from 0 to 9. Thus, B0 is the first member of Class B spectra and B9 is the last.

Class	Main Spectral Lines
O	Ionized helium, nitrogen, oxygen and silicon; hydrogen weak.
B	Neutral hydrogen and helium; ionized oxygen and silicon; ionized helium absent.
A	Hydrogen strong, ionized magnesium and silicon; ionized calcium, iron and titanium begin to appear; helium absent.
F	Ionized calcium (Ca II) strong; some ionized and neutral metal atoms (iron, manganese, chromium, etc.); hydrogen weak.
G	Ionized calcium strong; neutral metal atoms increasing and ionized forms decreasing; molecular bands of CH and CN appear.
K	Neutral metal atoms (including calcium) strong; molecular bands stronger; hydrogen very weak or absent.
M	Neutral metal atoms very strong; TiO bands appear.
R-N	Similar to K and M, but molecular bands of CH, CN and C_2 strong; TiO absent.
S	Neutral metal atoms strong; oxide (ZrO, LaO and YO) bands strong.

Stellar Temperature

Most stars have surface temperatures between 5,000-72,000 F. degrees and fall within the Draper classification.

Ultraviolet data from OAO-A2 will more precisely pinpoint stellar temperatures of young stars. This will provide accurate estimates of their helium content so astronomers can learn more of the nuclear processing which has gone on in the recent past.

The color index (see next page) for the hottest stars (O and B) is negative. These stars are very bright and are blue or blue-white in color.

The F stars are white and the G stars, like the Sun (G2), are yellow.

Cooler stars (K and M) have positive color indexes.
Their colors are orange and red.

Spectral Class and Surface Temperature of Main-Sequence Stars

Spectral Class	Temperature (F.)	Color Index
O5	72,000	-
B0	52,000	Minus 0.32
B5	30,000	Minus 0.16
A0	17,800	0.00
A5	15,500	Plus 0.15
F0	13,500	Plus 0.30
F5	12,000	Plus 0.44
G0	10,900	Plus 0.60
G5	10,200	Plus 0.68
K0	9,100	Plus 0.82
K5	7,800	Plus 1.18
M0	6,700	Plus 1.45

Their Ages

Ultraviolet measurements from OAO-A2 will pinpoint the
age of stars more accurately than is possible through visible
observations from ground-based observatories.

Telescopes on OAO-A2 will determine the mass of stars,
and the rate that the mass is being converted into energy.
These two ingredients are required for determining age.

Stars are said to be "zero age" when they reach the main
sequence. Their actual age is somewhat greater.

The Sun in our solar system is estimated to be "middle age," about 5 **billion** years old. It has burned up about five per cent of its hydrogen.

Assuming the Sun continues to release energy at the same rate, it will be about 10-billion years old before it departs from the main sequence and changes its character.

During the same period, stars with solar masses 10 times more than the Sun will depart from the main sequence in about 10 million years. The greater the stellar mass, the shorter life expectancy it has.

Pet Stars

Three pet stars which the Wisconsin experimenters are anxious to study are Betelgeuse, Gamma Velorum and Sigma Orionis.

Betelgeuse (M-type star) is very large (1,000 times greater mass than the Sun), very cool and is rapidly losing mass. The Wisconsin experiment probably won't be able to see the actual star, just the outer shell (chromosphere).

Gamma Velorum, W-R star (named after two astronomers, Wolf and Rayet, is shooting out materials at a very high velocity, some 3-million miles per hour. It is possible that these materials may be a source for X-ray stars. Several thousand volt protons are being cast off by Gamma Velorum.

Sigma Orionis E (early B-type star), some 1,500 light years from Earth, has a peculiar chemical composition which intrigues astronomers.

It has more helium than usual, but the exact amount is not known. Astronomers would like to know if Sigma Orionis E was helium rich to begin with or could it have come from evolutionary means.

NASA-INDUSTRY TEAM

NASA Headquarters

Dr. John E. Naugle   Associate Administrator for Space Science & Applications

Jesse L. Mitchell   Director, Physics & Astronomy Programs, OSSA

Dr. Nancy G. Roman   Chief of Astronomy, OSSA

C. Dixon Ashworth   Program Manager, Astronomical Observatories

T. B. Norris   Atlas-Centaur Program Manager

Goddard Space Flight Center

Dr. John F. Clark   Director

Joseph Purcell   OAO Project Manager

Dr. James E. Kupperian, Jr.   OAO Project Scientist

Kennedy Space Center

Dr. Kurt H. Debus   Director

Robert H. Grey   Director, Unmanned Launch Operations

John D. Gossett   Manager of Centaur Operations

Lewis Research Center

W. Russ Dunbar   Centaur Project Manager

Joseph A. Ziemianski   OAO Mission Engineer

Experimenters

Dr. Arthur D. Code   Chairman, Division of Astronomy University of Wisconsin, Madison

Dr. Fred L. Whipple   Director, Smithsonian Astrophysical Observatory, Cambridge, Mass.

Prime Contractors

Dean Davis
    Centaur Program Manager,
    General Dynamics-Convair,
    San Diego

Nicholas S. Snider
    OAO Program Director
    Grumman Aircraft Engineering
    Corp., Bethpage, N. Y.

Industry Team

Prime Contractors	Responsibility
Grumman Aircraft Engineering Corp., Bethpage, N.Y.	OAO-A2 spacecraft
General Dynamics-Convair San Diego	Atlas-Centaur Launch Vehicle

Experiments

Cook Electric Co. Morton Grove, Ill.	University of Wisconsin Experiment
Electro-Mechanical Research Inc., Princeton, N.J.	Celescope (Smithsonian Experiment)

Major Spacecraft Subcontractors

Company	Responsibility
AVCO Corp. Electronics Division, Cincinnati	Command Receiver
Bendix Corp. Electric Power Division Eatontown, N. J.	Power Regulation and Control Units
Bendix Corp. Navigation & Control Division Teterboro, N. J.	Reaction Wheels
Dalmo Victor Co. Belmont, Calif.	Magnetic Unloading System
Dorne & Margolin, Inc. Bohemia, N. Y.	Diplexer and Hybrid Junction
General Electric Co. Space Systems Operations Valley Forge, Pa.	Stabilization and Control System

Gulton Industries Alkaline Battery Division Metuchen, N. J.	Batteries
Gulton Industries Engineered Magnetics Division Hawthorne, Calif.	Voltage Inverter-Regulator Converter
Hazeltine Electronics Division Little Neck, N. Y.	Command Control Junction Box
Hughes Aircraft Co. Culver City, Calif.	Solid State Transmitters
IBM, Federal Systems Division Owego, N. Y.	Primary Processor and Data Storage
ITT Federal Laboratories San Fernando, Calif.	Boresight Star Tracker
Kollsman Instrument Corp. Syossett, N. Y.	Star Trackers
Textron Electronics Inc. Spectrolab Division Sylmar, Calif.	Solar Cell Arrays
Radiation Inc. Melbourne, Fla.	Experiment and Spacecraft Data Handling Equipment

-end-

America's
Finest
Rockets™

MOONCAT® **COLLECTIBLES SAN DIEGO, CA**